Post-Humanitarianism

Post-Humanitarianism
Governing Precarity in the Digital World

Mark Duffield

polity

Copyright © Mark Duffield 2019

The right of Mark Duffield to be identified as Author of this Work has been asserted in accordance with the UK Copyright, Designs and Patents Act 1988.

First published in 2019 by Polity Press

Polity Press
65 Bridge Street
Cambridge CB2 1UR, UK

Polity Press
101 Station Landing
Suite 300
Medford, MA 02155, USA

All rights reserved. Except for the quotation of short passages for the purpose of criticism and review, no part of this publication may be reproduced, stored in a retrieval system or transmitted, in any form or by any means, electronic, mechanical, photocopying, recording or otherwise, without the prior permission of the publisher.

ISBN-13: 978-0-7456-9858-8
ISBN-13: 978-0-7456-9859-5 (pb)

A catalogue record for this book is available from the British Library.

Library of Congress Cataloging-in-Publication Data
Names: Duffield, Mark R., author.
Title: Post-humanitarianism : governing precarity in the digital world / Mark Duffield.
Description: Medford, MA : Polity, 2018. | Includes bibliographical references and index.
Identifiers: LCCN 2018010023 (print) | LCCN 2018025557 (ebook) | ISBN 9780745698625 (Epub) | ISBN 9780745698588 (hardback) | ISBN 9780745698595 (pbk.)
Subjects: LCSH: Humanitarian intervention--Developing countries. | Humanitarianism. | Technology--Social aspects--Developing countries. | Developing countries--Social conditions--21st century. | Developing countries--Economic conditions--21st century.
Classification: LCC JZ6369 (ebook) | LCC JZ6369 .D84 2018 (print) | DDC 341.584091724--dc23
LC record available at https://lccn.loc.gov/2018010023

Typeset in 10.5 on 12pt Sabon by
Fakenham Prepress Solutions, Fakenham, Norfolk NR21 8NL
Printed and bound in the United Kingdom by Clays Ltd, Elcograph S.p.A.

The publisher has used its best endeavours to ensure that the URLs for external websites referred to in this book are correct and active at the time of going to press. However, the publisher has no responsibility for the websites and can make no guarantee that a site will remain live or that the content is or will remain appropriate.

Every effort has been made to trace all copyright holders, but if any have been inadvertently overlooked the publisher will be pleased to include any necessary credits in any subsequent reprint or edition.

For further information on Polity, visit our website:
politybooks.com

For
Elliott & Rupert
A luta continua

CONTENTS

Preface		*page* viii
Chapter 1	Introduction: Questioning Connectivity	1
Chapter 2	Against Hierarchy	15
Chapter 3	Entropic Barbarism	28
Chapter 4	Being There	39
Chapter 5	Fantastic Invasion	57
Chapter 6	Livelihood Regime	79
Chapter 7	Instilling Remoteness	93
Chapter 8	Edge of Catastrophe	114
Chapter 9	Connecting Precarity	129
Chapter 10	Post-humanitarianism	143
Chapter 11	Living Wild	159
Chapter 12	Conclusion: Automating Precarity	178
Notes		193
Bibliography		202
Index		226

PREFACE

In the summer of 2012, I made a chance discovery, from which this book has grown. An internet search focusing on humanitarian work in Sudan returned some unexpected results. Since 2003, the UN had been working with a number of geospatial research institutes on the feasibility of using high-resolution satellite imagery to help the emergency response in Darfur. Apart from mapping refugee camps, this included remotely finding water sources so that new camps could be located nearby. Particularly striking, however, was the development of algorithms able to estimate the varying size of these settlements almost as they changed, without the need to go there. Moreover, advocacy groups were already using satellite imagery in the usually dangerous work of documenting sites of possible human rights abuse, by identifying things like burnt buildings or disturbed ground safely from the air.

Having visited Sudan many times for work and research purposes, this remote sensing was new to me. For a while, the discovery conveyed an excitement akin to having stumbled into a parallel scientific universe. Reflecting the widespread belief among humanitarian agencies that aid work had become more dangerous, my research at the time focused on the growing retreat of international aid workers into fortified aid compounds and gated-complexes. Given this withdrawal, the significance of remote sensing was immediately apparent – that is, the potential of being able to understand and act from a distance without the need to be entangled on the ground. This ability is the central theme of *Post-Humanitarianism: Governing Precarity in the Digital World*. It explores a double movement, whereby a loss is also a gain. What is given up is the textured familiarity that disappears with remoteness. What is gained are the new visualization techniques, smart sense-making tools and novel governmental possibilities resulting from the digital recoupment of distance.

This discovery sparked a sustained period of enquiry, deepening interest and growing concern around the issue of remoteness and recoupment. There is more to becoming distant in order to get closer, so to speak, than meets the eye. Darfur quickly proved to be a small demonstration of that much wider event known as the computational turn – that is, from the mid-1990s, the rapid spread of computers, mobile devices, software platforms and social media into all aspects of personal, public and international life, without exception. Moreover, as the retreat of aid workers suggests, this veritable electronic globalization gathered momentum at the same time as the world was being seen as more unpredictable and dangerous than before. For most people, the two are unconnected. Indeed, given the often-claimed growth in political complexity and uncertainty, the new sense-making tools appear fortuitous. Sceptical of such coincidences, *Post-Humanitarianism* is more open to their formative interleaving and interdependence.

The ideas of remoteness and recoupment beg the question of what has been lost, and what exactly has been gained? The computational turn has fundamentally changed our understanding of the world and what it means to be human. Together with a loss of familiarity, a growing reliance on machine-thinking means that reason, human agency and being present in the world no longer occupy the valued place they once did. For much of the twentieth century, human behaviour was measured against the often untrustworthy, but rational, decision-maker *Homo economicus*. Displacing this once familiar liberal avatar, a new behavioural yardstick has appeared.  Compared to the confident *Homo economicus*, this new figure is unsure of itself. Cognitively challenged by the speed and complexity of contemporary life, it relies more on automatic and unconscious sense-making mechanisms rather than on conscious rational thought. This post-human avatar struggles to take in more than the immediate givens of its environment and accessible networks. The pages of this book are haunted by this overwhelmed, distracted and necessarily ignorant subject that, for want of a better term, is here dubbed *Homo inscius*. While challenged by a complex and uncertain world, *Homo inscius* is comfortable with its own inner vulnerabilities and human limitations, and its consequent dependence on the compensatory attentiveness that smart technology affords.

Rather than accept this rather unappealing subject as the guide to our purported post-human future, this book has sought perspective through a return to the materiality of capitalism. As an object of study and critique, capitalism has been long neglected within the academy. The computational turn is an intrinsic part of the transition from the mass production of Fordism to the personal consumption of today's new economy where *Homo inscius* is its ideal customer and pampered savant.

When this work began, it was unforeseen that it would demand a reinterpretation of my previous research and experience. After decades in the post-structural wilderness, I was emboldened to return to the structuralism of my Marxist youth. Earlier work on global governance, disasters and security, once felt to be outside and critical of the status quo, now feels inadequate. Not least, the crisis has deepened. With its alienation, patriarchy and racism, the past was never perfect. Since the 1960s, when these oppressions and insults were forcefully identified, a limited institutional progress has been made. The original aim, however, had been to change society radically as a whole for everyone. This revolutionary impulse has stalled and been deflected. Today's world, moreover, is different. Those areas of educational, economic and political autonomy that once supported independent circulation, allowed resistance and encouraged utopian visions of a better life have narrowed and disappeared.

While contestation and protest continue, much of this energy has been captured and set to work on adapting and deepening the system rather than fundamentally changing it. The problem is that, while reform may have worked in the past, the dots no longer join up. For those vast and ever growing numbers of people obliged to live on the edge of disaster, a future of precarity in an automating and polarizing world is calling. Capitalism, moreover, is being adapted to realize and profit from such a techno-barbaric future. Pitched against this violence-edged intent, however, new and diverse areas of autonomy and forms of recalcitrance are struggling to emerge. In this dangerous interregnum, speculative theory, searching critique and, not least, a solidarity born of political faith in humanity are needed more than ever.

Writing a book is as much a lived relationship as an academic exercise. Acknowledging one's debts is always incomplete. Over the last few years, they have been many, diverse and sometimes unexpected. In taking stock, a warm thanks to Finn Stepputat and his colleagues at the Danish Institute of International Studies (DIIS) in Copenhagen. In the spring of 2013, I had the pleasure of being a guest professor at DIIS. This provided an opportunity to embark properly on the preparatory work for this book. Many thanks also to Luis Lobo-Guerrero for kindly arranging a similar visit at the end of 2014 to the department of International Relations at the University of Groningen. Over the past few years, I have presented the evolving themes of *Post-Humanitarianism* in a dozen workshops and conferences in Britain, the Netherlands, Germany and Lebanon. As such occasions are useful for testing ideas and getting audience feedback, I would like to thank their organizers. In particular, I am grateful to Jolle Demmers in the Centre for Conflict Studies at the University of Utrecht for her interest and support.

At the beginning of 2014, I returned to Sudan to revisit the village where I completed my Ph.D. fieldwork in the mid-1970s. This was invaluable in helping me understand the changing nature of international space. For their advice, help and friendship, I must thank Ahmed Gamal Eldin, Al-Amin Abumanga, George Pagoulatos and Osman el-Kheir, together with Babiker Osman Shanowa, Fiesal Kaghu, Mohammed Osman, Sameer al-Tayeb al-Shami and the rest of my friends in Maiurno. Kind mention must also be made of the many people who have in some way inspired, helped or been tolerant. These include Adbullahi Gallab, Alison Howell, Antonio Donini, Bertrand Taithe, Colleen Bell, Dan Large, David Turton, Diana Felix da Costa, Jan Bachmann, Jonathan Fisher, Judith Squires, Juliano Fiori, Martin Gainsborough, Mathew Bywater, Nada Ghandour-Demiri, Nick Stockton, Norah Nihland, Oliver Richmond, Orit Halpern, Roger Mac Ginty, Sara Pantuliano, Sarah Collinson, Sophia Hoffman, Thea Hilhorst, Tim Edmunds, Tom Scott-Smith and Zoe Marriage. Given my frequent requests for deadline extensions, I am grateful to Louise Knight and Nekane Tanaka Galdos at Polity Press for their patient support. I hope the wait has been worth it.

My friends and academic comrades Jens Sorensen and Ray Bush deserve special thanks. It's always great to meet and talk things through. Many thanks also to Vanessa and Mladan Pupavac for opening my eyes to literary critique and the realities of the migration crisis. Douglas Spencer's work on neoliberalism and architecture has been inspirational. For the changing regimes of international aid, Susanne Jaspars' pioneering research has been immensely helpful. In terms of how we understand the world and our place within it, Brad Evans, David Chandler and Julian Reid continue to challenge conventions and set the pace. Not least, I owe Rupert Alcock a huge debt of gratitude for kindly reading the first draft of the book. Besides his own path-breaking work on cognitive politics, his suggestions and impressions have been invaluable. Any mistakes or shortcomings are, of course, my own. I have learnt a lot and thank them all for their conviviality and kind consideration.

Finally, at a more personal level, I'd like to thank Dr Tahir Shah and his wonderful team at the Queen Elizabeth Hospital, Birmingham, for their warmth and professional care. I would also like to mention Geoff Whitehouse, George Cree, Chris Lloyd and Graham Ward for their enduring friendship, humour and good company. I am similarly indebted to 1st Sedgley Morris Men for taking me under their wing. My life would be diminished without their camaraderie, mischief and laughter. Last but not least, in what has often felt to be a never-ending task, I have enjoyed the love and support of my family and my wife Jean. This book would otherwise never have been completed.

Mark Duffield
Sedgley
January 2018

Chapter 1

INTRODUCTION: QUESTIONING CONNECTIVITY

At the time of writing, there is a consensus among Western security specialists that the world has entered a period of uncertainty and political instability unprecedented in recent times. One such source is the latest Munich Security Report (MSR 2017) provocatively entitled 'Post-truth, post-West, post-order?'. Intended for policy and security professionals, the Report is a digest of the latest international trends and events. Like a breathless messenger, it describes the different flags and factions of the illiberal barbarians now massing at the gates. In concert with a clutch of new books,[1] it depicts a groundswell of populist and fundamentalist movements, laying claim to local or cultural authenticities, which are now challenging and pushing back cosmopolitan values and libertarian identities. Expected since the mid-1990s, it looks as if the 'coming anarchy' may now be arriving (Kaplan 1994). There are several factors, however, that give the present a new and distinct feel. Divisions and contradictions are appearing in the West. Random terrorism is becoming routine, while dissatisfaction is growing among those who feel left behind and abandoned. Apart from increasing security measures and orchestrating public displays of resilience, political elites are challenged for real answers. With Syria as a case in point, compared to the 1990s, Western states have also lost their interventionary nerve.

Citizens of democracies believe less and less that their systems are able to deliver positive outcomes for them, and increasingly favour national solutions and closed borders over globalism and openness. Illiberal regimes, on the other hand, seem to be on solid footing and act with assertiveness, while the willingness and ability of Western democracies to shape international affairs and to defend the rules-based liberal order are declining (MSR 2017: 5).

This book is not concerned with questioning whether this picture of international push-back and Western decline is accurate or not. That it exists and has credence is sufficient. Our point of departure is the stark contrast between this imaginary future–present and a different, earlier one – namely, how the international scene looked a mere five or six decades ago. Driven by frequently violent struggles for national liberation, decolonization and the dismantling of imperialism from below were in full swing. With its excess of youthful radicalism, for many commentators the 1960s were a volatile interregnum of emancipatory forces pushing towards world revolution (Mills 1960). Breaking with Victorian Marxism, the rash of anticolonial struggles ushered in a New Left convinced that the peasantry was now the true heir of this revolution. As the colonial order eroded, continuing privation and exploitation meant that it was the peasantry, unlike most industrial workers, that now had nothing to gain from compromise: 'In China and Vietnam, in Cuba, Kenya and Algeria, in Brazil's North-east and in the back-country of Angola, the peasantry has emerged as the decisive force in revolutionary struggles' (Buchanan 1963: 11).

Contrary to an earlier Eurocentric left orthodoxy, while a radicalized intelligentsia and worker vanguard could prime the revolutionary fuse in the industrial countries, it was an emergent Third World that would now ignite it (Marcuse 1967). Moreover, without the active alignment and international solidarity between these spatially separated forces and struggles, the chance of world revolution would be lost. Whether such views were realistic or delusional should not detract from the fact that they were real enough to mobilize people on an international scale. The contrast between a revolutionary, anti-racist future–present, where the international appeared as a space of political optimism and fraternity, and today's more pessimistic vista of rupture and political failure is striking.

This book is a preliminary attempt to try to understand this shift and assess what we may have lost and, for good or ill, what we have gained. Methodologically attentive to history, it addresses this question in relation to the changing understanding of the nature of humanitarian disaster. How disasters are understood and communicated shapes the nature of the global North–South interface (Chouliaraki 2013).[2] Indeed, one could go further. Since the 1980s, disasters have become a new ontological force. From the crash of asteroids into a primeval Earth, disasters have been given a pivotal role in the evolution of life, in the development of creativity and, not least, as key punctuation marks in the emergence and spread of human society (Homer-Dixon 2007). This catastrophism has accompanied the rise to dominance of an ecology-based resilience thinking, with its signature view that 'authentic' life exists in the *jouissance* that lies on the edge of extinction. Resilience is a measure of the probability of escaping disaster through socializing the smart moves that drive developmental evolution (Holling 1973). Disasters are thus a potent bridging mechanism that connects humanitarian practice with wider ideological and societal change. These changes, moreover, help illuminate the move from optimism to political pessimism. This shift, it will be argued, is integral to the rise of post-humanitarianism.

However, in making a link from disasters to these broader questions, two additional and accompanying registers or sets of differences are important. Over the period in question, there has been a spatial shift from 'circulation' to 'connectivity', together with an interrelated ontological, epistemological and methodological transition from deductive 'knowledge', framed by history and causation, to an increasing reliance on inductive mathematical 'data' and machine-thinking for sense-making. The way we know the world and understand what it means to be human has fundamentally changed (Chandler 2018). Rather than seeing the emergence of a new post-human essence, this book grounds these shifts and registers in the changing nature of capitalism. While corporations, governments and the academy celebrate the age of connectivity, and regard the sort of international foreboding described in the Munich Security Report as a separate issue, we are more open to the possibility of their causal correlation. This Introduction unpacks these registers and gives the reader an indication of the structure of the book.

Circulation and Connectivity

Between the 1960s and the present, the nature and organization of international space have changed. Of primary importance has been the relative shift from 'circulation' to 'connectivity' (Reid 2009). As a factor of spatial organization, circulation involves the physical movement or flow of people and things within, across or around terrestrial milieus and topographies. Discussed more fully in chapter 5, Foucault has argued, that the principle of circulation was central to a liberal conception of security arising from the discovery of the early modern town in terms of its spatial and logistical dynamics. The problem of the town 'was essentially and fundamentally a problem of circulation' (Foucault 2007: 13). During the nineteenth century, improving the circulation of people, goods, sewage, light and air, together with managing the movement of disease, crime and political unrest, would become a key feature of modernist planning and urban design (Rabinow 1995). From the perspective of modern urban planning, the city was an infrastructure designed to maximize the circulatory potential of autonomous people and things, while controlling the bad and inimical. Through the opening-up achieved by roads, canals, sewers and railways, for example, people and things were enabled to move, change place and transact. While not without risks, and thus needing administrative, health and police oversight, the aim was to maximize circulation along such fixed conduits.

Connectivity is similar but fundamentally different. Google's notion of a data-based urbanism, for example, sees cities as key sites for the conversion of data extracted from the electronic interactions of individuals into continually adapting forms of artificial urban intelligence. A 12-acre site in Toronto's waterfront area is currently being developed as a testbed. It envisions: 'Modular buildings assembled quickly; sensors monitoring air quality; traffic lights prioritising pedestrians and cyclists; parking systems directing cars to available slots; delivery robots; advanced energy grids; automated waste sorting and self-driving cars' (Morozov 2017).

Here the city appears as a closed interactive milieu involving the continuous recording and exchange of information between people, things and computer interfaces in motion. Connectivity draws together different domains such as consumer needs, waste disposal, transport, parking and delivery requirements into an integrated real-time information network. While people and things still move, change place and transact, it is no longer autonomous circulation in the modernist sense. Without triggering a series of alerts, a person could not, for example, arrive unexpectedly at a railway station, and buy a ticket for destination A but leave instead at station B. Within the smart city, movement and behaviour are constantly recorded, algorithmically analysed, optimized and directed (Halpern 2014b).

Unlike the spontaneous circulation allowed by the modern city, movement within the smart city is essentially robotic.

As a science of information, cybernetics requires the recording and storing of data on all past interactions as a precondition for predicting future behaviour and signalling the presence of anomalies (Wiener 1954). Unlike free circulation, which always involves a potential threat to security (Foucault 2007: 19), connectivity uses the command and control functions made possible by data informatics to avoid surprise. To put this another way, while circulation is necessary it is also open to accidents, dangers and unforeseen consequences. Air travel, for example, can be a vector in the spread of disease. As a way of controlling the necessary risks of circulation, security has evolved as an expanding and invasive technology of connectivity (see chapter 5).

There is another aspect of connectivity, however, that is also important for this book, and which further distinguishes it from the territorially grounded nature of circulation. Imagine a dozen computers scattered around the globe, networked together via a central hub and each machine being able to transmit and exchange data with the others instantaneously. Rather than having to flow through or circulate within frictive topographies, connectivity has the power to leap directly across them, bypassing terrestrial insecurity while rendering distance insignificant. Finance capital, for example, is not like physical money. The latter constantly circulates between pockets, cash registers and banks until it is worn out. As an example of connectivity, finance is capital encoded as data that travels at the speed of light between the vast territorially dispersed network of computers that constitute the global banking system (Lewis 2014): '[Connectivity] de-spatializes the real globe, replacing the curved earth with an almost extensionless point, or a network of intersection points and lines that amount to nothing other than connections between two computers any given distance apart' (Sloterdijk 2013 [2005]: 13).

Although different, circulation and connectivity are not mutually exclusive. They exist together, shape each other and, over time, exist in varying combinations. For this book, the relative shift from circulation to connectivity is implicated in the displacement of revolutionary optimism by political pessimism. In the 1960s, at the height of international expectation, the ability for people, their histories, experience and politics, to circulate internationally was greater than it is today. For a while, the circulation and flow of political praxis was possible as never before. During the period of decolonization, Western European countries were moved to accept permanent immigrants from their colonies and former colonies, together with allowing refugee settlement and recruiting significant numbers of migrant workers. Aspirational white settler colonies such as Australia, New Zealand and Canada also temporally lifted the 'colour line' that had earlier applied, especially toward Asian labour migrants (Meyers 2002). For Herbert Marcuse, as for other radicals exiled at some point in their lives, the ability for political praxis to circulate was taken for granted. At a time when journalists were not embedded (Page 1989), this ability was an essential condition of the international solidarity necessary for world revolution. By the mid-1970s, however, the near-universal curtailment of immigration was already underway.

Driven by a mix of racial, social and security fears, the relative post-World War II openness to migration has narrowed and closed under successive waves of immigration controls, nationality laws and refugee restrictions (Hammerstad 2014). Since the end of the Cold War, as a visible register of this institutional move to closure and return, the number of physical border fences, demarcation walls or separation zones to contain the risk of autonomous movement has exploded globally (Brown 2010). Of course, the barriers and restrictions that now striate the globe have not prevented the urge to move. Indeed, as the upward track of numbers suggests (UNHCR 2017a), the pressure to escape poverty, disaster and war, even at the risk of an arduous and perilous passage, is as strong as ever. With millions in the queue, it shows few signs of abating. While offering no viable solution, the interdiction and return measures used to insulate the West have done little more than criminalize autonomous human circulation.

Connectivity and remoteness

As the legal circulation of migrants, refugees and other *sans-papiers* has narrowed and closed, in terms of the data being stored and exchanged between machines and screen interfaces, connectivity has expanded exponentially (Cortada 2012). At the same time, computational technologies including remote satellite sensing, computer modelling and Big Data informatics have come to shape a dominant, if particular, understanding of the world, how it works and the status of the humans that inhabit it (Halpern 2014a; Chandler 2018). Climate change, for example, was a key discovery of predictive computer modelling (Edwards 2010). The juxtaposition between the international closure to the circulation of political praxis and the expansion of data connectivity and its new remote sense-making tools is a formative tension that runs throughout this book. To put this another way, since the 1990s there has been an associated growth in physical and existential 'remoteness' from the world that is being compensated by the digital recoupment of distance. Remoteness, however, is ambiguous. It is negative, as in a loss of familiarity, while also being a positive condition – that is, as a challenge for technoscience to overcome.

A negative remoteness is not only reflected in the erection of physical and technological barriers to stop the circulation of political praxis; it can be seen at many levels, including the fragmentation of nations. With examples spanning the globe, over the last three or four decades many erstwhile multicultural or mixed societies have been wrenched apart, fragmenting and polarizing along inimical ethnic, cultural and religious lines (Gregory 2008; Sorensen 2014; Mishra 2017a). Mid-level technological societies have been reduced to – or, should we say, 'revealed' as – a chimera of competing tribal amalgams (Usborne 2004). As if designed for it, the trend towards individuation, separation and polarization has taken to social media with alacrity (McBain 2014; O'Callaghan et al. 2014; Cadwalladr 2017). As discussed in chapter 7, through a combination of risk aversion and political push-back, a loss of familiarity can also be seen in the increasing absence of grounded international aid workers, journalists and academics within 'challenging environments' (Healy & Tiller 2014). President Trump's travel ban on selected Muslim countries, and the current uncertainty over the future of EU nationals in Brexit Britain, are symptoms of this pervasive, and often violent and discriminatory, tendency towards distancing and a loss of familiarity.

Remoteness, however, also has a positive dynamic that springs from the ability of connectivity to leap across, sidestep or pass beneath the ground friction[3] of a dangerous world productively, while simultaneously creating new ways of knowing and appropriating that world. First identified over fifty years ago, the inverse relationship that technoscience establishes between familiarity and distance is what Hannah Arendt called 'world alienation' (Arendt

1998 [1958]: 48–254). The paradox of exploration is that, while its aim was to widen horizons, the maps and charts of the early modern age 'anticipated the technical inventions through which all earthly space has become small and close at hand' (1998 [1958]: 251). This shrinking of the globe has continued through the surveying capacity of the human mind, 'whose uses of numbers, symbols, and models condense and scale earthly physical distance down to the size of the human body's natural sense and understanding' (1998 [1958]: 251). The shrinkage of the Earth, however, has been compensated for by the objectivity that distance gives. Objectivity necessitates a disentanglement 'from all involvement in and concern with the close at hand' (1998 [1958]: 251). For Arendt in the 1950s, the decisive technology of shrinkage was the aeroplane. The advent of satellites, geospatial technology and interactive broadband, however, redoubles her point. The ability to leave the Earth, either physically or as an Internaut,[4] 'is like a symbol for the general phenomenon that any decrease of terrestrial distance can be won only at the price of putting a decisive distance between man and earth, of alienating man from his immediate earthly surroundings' (1998 [1958]: 251).

World alienation is the hallmark of the modern age and is 'inherent in the discovery and taking possession of the earth' (1998 [1958]: 254). As the political history of maps suggests (Wood 2010), remoteness and distance call forth new sense-making tools which furnish new ways to strategize and project power – and, thus, to appropriate and reappropriate the Earth.

Knowledge, data and post-humanitarianism

As a function of the reappropriation of the modernist legacy currently under way by the agents of the new economy (Boutang 2011 [2008]; Srnicek 2016), the recoupment of distance through digital connectivity has its own history of abstraction and violence.[5] The textured histories, motivations and justifications of distant or now-hard-to-reach people, once familiar through face-to-face exchange or the ethnographic encounter (see chapter 4), have been transformed for the convenience of mathematics into electronic data. To make behavioural patterns amenable to visual representation, knowledge has been reworked into digital signals and alerts able to be recorded and algorithmically analysed by machines (see chapter 5). As a tool for knowing and appropriating the world afresh, the supplanting of the grounded ontologies of circulation/knowledge with those of connectivity/data has not been frictionless or straightforward (Amoore 2011). Moreover, it has been far from natural, or a simple matter of technological change.

As world history suggests, knowledge can be put to many uses, including vile, repressive and genocidal ones. However, while murderous dictators may wish otherwise, knowledge is never closed to itself. Knowledge affords its own critique. Even the slaves of San Domingo could dream of freedom through the Enlightenment texts that their masters betrayed (James 2001 [1938]). Data is different from – even antagonistic to – knowledge (Galloway 2013; Chandler 2015). Knowledge is open to intentions, justifications and causes. It emerges from empirical experimentation, ethnographic encounters and deductive causal reasoning. By comparison data focuses on the potentialities of individuals as derived from the inductive statistical analysis of their past behaviour.

Knowledge admits to a distinction between the 'reality' of lived experience and an existing and structurally defined 'world' inhabited by actual, present and sentient subjects. The space, or commons, between reality and the world allows room for different truths, competing theory and critique (Rouvroy 2012). Knowledge is inseparable from the contested political commons producing it. Data allows no such distinction or space. Reality and the world are indistinct, and existence is a condition of the pure, unmediated factuality of virtual and probabilistic subjects. Whereas knowledge allows for consciousness, reflexivity and theorizing, data is more concerned with signals, alerts and reflexes, and the unconscious or unreasoned dimensions of human behaviour (World Bank 2015). As such, the very possibility of theory and reasoned critique is questioned (Anderson 2007).

Despite knowledge and data being antagonistic regarding how we understand the world and what it means to be human, the current hegemony of what Antoinette Rouvroy (2012) has called 'data behaviourism' in explaining social and environmental phenomena has largely escaped critical concern. If anything, its rise has been welcomed and seen as fortuitous given the 'complexity' of the problems we face. As discussed in chapter 4, this is partly explained by the dominance within the academy of empirical and behavioural modes of understanding and methodology. Included here are the various strands of work often drawn together as post-humanism

(Braidotti 2013). For example, the new empiricism, speculative realism and actor network theories that variously draw on process-oriented behavioural ontologies of becoming, in which materially embedded individuals are held to exist in an unmediated empirical relationship within their enfolding environments (see Galloway 2013; Chandler 2015). Like data behaviourism, the pure factuality of a post-human existence also casts doubt on the distinction that knowledge produces between reality and the world. Without this distinction, the world becomes smaller than the sum of its parts (Latour et al. 2012). An individual's mental horizon reduces to the immediate who, where and when of their changing network connections and disconnections.

Post-humanitarianism goes beyond the smart forms of humanitarian intervention and design that have emerged from the rubble of 1990s liberal interventionism following the West's foreign policy disasters in the Middle East. This book does not add to the growing pile of optimistic declarations of a fresh humanitarian start based on these technologies (Meier 2015). Post-humanitarianism is the international face of post-humanism. It gives Arendt's notion of world alienation through growing technoscientific remoteness a new dimension and meaning. At a time of deepening polarization, fragmentation and anger, the post-humanitarian turn to narrow empiricism, unmediated experience and data behaviourism as the international optic of choice is short-sighted, to say the least. Allowing no distinction between reality and the world, and asserting the design principle over any need for radical change, post-humanitarianism lacks any political, historical or moral perspective save that of its own importance. Yet it is also a positive and active force. Post-humanitarianism is central to capitalism's moving beyond the enclave or special economic zone to incorporate the vast informal economies of the global South – a move discussed in chapter 9. It is a key departure in fashioning the disaggregated biopolitical technologies necessary to support the social reproduction of an expanding global precariat in a post-social world (see chapters 11 and 12).

Boomerang effect

The emergence of post-humanitarianism is intimately bound up with the computational turn – that is, the steady penetration, since the 1980s, of computers, the internet, mobile telephony, interactive broadband, software platforms, social media and automating apps into all aspects of personal, social, national and international life *tout court*. Given the antagonistic relationship between the knowledge/circulation and data/connectivity registers, what is remarkable about the computational turn is its seamless, unremarkable and almost natural arrival. To paraphrase John Robert Seeley's reflection on the British Empire, we seem to have readily transferred to machines the ability to think on our behalf, in a fit of absent-mindedness. Yet, this naturalness is illusionary.

The arrival of the computational turn was preceded by a wide range of contributory and anticipatory developments. Before the commercial arrival of computers, the transformation of textured knowledge into mathematical data was already under way. These anticipatory developments prepared the ground, as it were, for the seamless datafication of society. As argued in this book, a strategic site for the transformation of knowledge into data, and the current refinement of the smart technologies that have emerged, is the global North–South interface. Moreover, since the 1980s, the changing understanding of what constitutes a humanitarian disaster has been instrumental (see chapter 5). The anticipatory and developmental role of disaster is a contemporary example of what Hannah Arendt called the 'boomerang effect' (Arendt 1994: 155; also Foucault 2003: 103).

For Arendt, the moral licence and lack of restraint that characterized nineteenth-century imperialism served as a trial run for the European totalitarian regimes of the twentieth century. Throughout the colonial period, the colonies functioned as general laboratories, testbeds or sites of anticipation for emerging capitalist relations and new modes of governance that would materialize in Europe. These ranged from prison reform through public health to centralized policing and modernist urban planning (Rose 2000: 107, n25; Rabinow 1995).[6] While the boomerang effect has a history, it also operates in the present. Relatively deregulated and with weak data protection laws, the global South continues to be a testbed for new technologies (Jacobsen 2015). The boomerang effect is a disruptive concept. It unsettles modernist ideas of developmental and temporal sequencing by according the global South an experimental or forward-looking role; the South is where our post-humanitarian future lies.

Structure of the Book

In seeking to understand the shift from optimism to political pessimism, this book examines how the reason and human agency associated with the former have been seamlessly transferred to the automatic devices and smart technologies that underpin the post-humanist turn. In analysing the arrival of the computational turn, seeing its roots in the positive or creative urge driving contemporary capitalism has been central. *Chapter 2* addresses this issue through the idea of a new spirit of capitalism. Ironically, this spirit drew much of its energy from the international revolutionary and countercultural upsurge that peaked during the late 1960s and early 1970s, and which is here called the May '68 critique. Recouped as a progressive neoliberalism, the resulting problematization of the alienation, hierarchy and patriarchy of welfare-Fordism played a formative role in the transition to a personalized, computer-based new economy. In the process, the new spirit has been willing to exchange the security of welfare-Fordism for the freedom of the market in a post-social world – that is, a world where social protection and state provision are attenuated or no longer exist. The idea of the post-social, its anticipation in the global South, and the social automation necessary to make it a reality, are essential constituents of post-humanitarianism.

During the 1960s, however, these outcomes were still uncertain. The postmodern age has been shaped by a questioning of the sovereignty of the subject from many competing viewpoints. *Chapter 3* looks at the contrasting positions of New Left Marxism and cybernetics on this issue. The former, through the negative dialectic, saw the possibility of capitalism's decent into a techno-barbarism unless prevented by revolution. For cybernetics, the problem was one of avoiding entropy by using the command and control functions made possible by a science of information able to govern human–machine interaction. In the transition to post-Fordism, it was cybernetics that triumphed. As *chapter 4* argues, the May '68 movement was the last time an autonomous intelligentsia was able to make a stand against the rising tide of empiricism and behaviourism within the academy. To provide a comparison with the isolating risk aversion of the present, the chapter concludes with a review of the structural method and immersive anthropological fieldwork that were still possible in the mid-1970s.

As examples of the boomerang effect, *chapters 5* and 6 present non-governmental organizations (NGOs) and the changing nature of international aid during the 1980s as a preparatory stage for the computational turn and appearance of post-humanitarianism. In relation to the present, it was a phase of both anticipation and rupture. A new ontology of disaster emerged. Rather than a modernist separation from society, disaster became a defining characteristic of society and its inner vulnerabilities and weaknesses. The earlier social and political views of famine gave way to complexity thinking and indeterminacy. Experimentation with early warning transformed famines into the signals and alerts thrown off by behavioural change. Rather than theory, the new emphasis was operationality. These developments exemplified the coming of age of an essentially cybernetic understanding of liberal security. In anticipating the post-social, the fantastic NGO invasion also piloted the projectized forms of livelihood support and community development. While anticipatory, however, the project form, together with the direct humanitarian action of the period, were still vested in the primacy of human agency and grounded engagement.

With liberal interventionism now buried in the ruins of Iraq and Afghanistan, *chapter 7* examines the West's growing inhibition and retreat from the world through the culture and architecture of the fortified aid compound. Field-security training is used as an example of how the shift from circulation to connectivity is not natural, as it were, but must be taught. Training instils remoteness and how to view the world in post-human terms. Rather than a product of history and causation, the outside is now an unmediated flux of green, amber and red behavioural cues and environmental alerts. Resilience training reinforces a refocusing on the inner-self. The fortified aid compound functions as a therapeutic structure offering refuge and mental respite from an uncertain world that is no longer fully understood. While a cultural dead-end, the fortified aid compound provides a counterpoint to the liberating leap that has been made into the electronic atmosphere – the last global strategic plane where one-sided economic, political and cultural action is still possible.

The creation of a global precariat has been the single most significant achievement of capitalism's new economy. Precarity is also the governmental object of post-humanitarianism. *Chapter 8* examines the blurring of economy and disaster that brought the precariat into being. Made possible by the spread of mobile connectivity throughout the global South, new ways of valorizing the social reproduction of the precariat have emerged. As argued in *chapter 9*, smart technology folds downwards into the inequalities and mobility differences encountered. It has enabled capitalism to move beyond the confines of the special economic zone to enrol the precariat within the circuits of the global economy.

The remaining chapters focus on the new distance-recouping and sense-making tools that have emerged as part of the computational turn and form the technological basis of post-humanitarianism. *Chapter 10* examines remote satellite sensing and how refugees are now understood ecologically – that is, as part of the environments in which they are embedded. Crisis informatics has rediscovered disaster events as distributed information systems, and enabled, among other things, the appearance of a new generation of hyper-bunkered post-humanitarians. *Chapter 11* takes this analysis further by focusing on the disaggregated biopolitics of the biohuman as reflected in the notion of humanitarian innovation. This includes the emergence of a suite of attentive self-acting objects and smart technologies that have now absorbed the individual and collective human agency associated with earlier direct humanitarian action. These technologies are celebrated as enabling the precariat to survive in a post-social world, a key feature of which is the absence of a fixed infrastructural grid.

In drawing the book to a close, the *Conclusion* discusses the post-humanitarian attempts at streamlining and automating social reproduction among the precariat through cognitive science. Such developments dovetail with the long postmodern trope of a caretaker society – that is, having solved all important social and political problems, all that is now needed is piecemeal technological change. Such elite complacency, however, sits awkwardly with the paradox of connectivity – that is, the more connectivity, the greater the ground friction and political anger generated.

Chapter 2

AGAINST HIERARCHY

By the mid-1990s, the claim that we now live in a network society – indeed, in an information-centric network world – was already widely accepted (Castells 1996). A key economic change supporting the network metaphor was the reorganization of the Fordist company into a global brand. It should be emphasized that these changes would have been difficult without the extraordinary diffusion and global embedding of digital technologies. From the mid-1970s, with their decreasing size and expense, computers spread from large companies to embed in the national economic auditing systems of the global North. During the 1990s, via PCs, the internet and, eventually, smart portable devices, computers moved from the business sector and government departments into the homes and hands of Northern consumers. For the digital historian James Cortada, the end of the 1990s marks the completion of the first stage of computer diffusion (Cortada 2012). By this time, capitalism's new post-Fordist economy can be seen as having been reconstituted as a globally distributed information system. This aspect of second-wave globalization,[1] called 'deindustrialization' at the time, saw large-scale factory closures and redundancy in the industrial heartlands of Europe and the USA as manufacturing functions typically moved to the expanding low-wage enclaves of East Asia (Amsden 1990). Headquarter companies in the global North retained core activities, usually involving logistics, finance and design skills, while subcontracting or franchising manufacturing or service functions to a network of overseas ancillary companies and sites.

Given this pattern of global restructuring, this book uses the descriptive term 'network capitalism' interchangeably with 'new economy'.[2] The term 'network capitalism' is useful because it makes a structural and spatial link with connectivity at the same time as anticipating the increasing datafication of capitalism's global logistical network. According to Cortada (2012), the second and present stage of digital embedding began towards the end of the 1990s before accelerating from the mid-2000s. It is based on interactive broadband and the rapid expansion of geolocational mobile telephony as the commercial interface of choice. The main feature of this second stage has been the rapid leapfrogging of broadband and mobile telephony into the extensive informal milieus of the global South. The implications of this are discussed in chapter 9. Together with the financialization and liquefaction of capital, this embedding and digitalization has been a necessary condition for the global expansion of the complex business logistics that underpin network capitalism's dispersed just-in-time manufacturing, assembly and delivery systems (Cowen 2014). Lying at the heart of network capitalism, a multilevelled web of financial circuits and global supply chains interconnects Europe and the USA with East Asia and the global South. Reliant on the power of connectivity to leap across the intervening barriers, inequalities and expropriations – indeed, to occlude such violence and ground friction – this vast network of global logistical systems has enabled the replacement of an earlier culture of mass consumption in the global North by post-Fordist patterns of personalized pampering.

This chapter continues to introduce concepts and ideas that are important for this book. Its main purpose, however, is to outline some of the cultural and political forces involved in the transition from Fordism to network capitalism. In particular, those forces driving the boomerang effect, or feedback loop, connecting the global North and South during this transitional period. Originating in the May '68 anti-capitalist critique, a new cosmopolitan spirit of capitalism emerged. For authenticity and creativity, while against hierarchy and patriarchy, this spirit shaped the emergence of the new economy. A key feature of the time was the exchange of the normative and social insurance-based security of welfare-Fordism for the individual freedom of the market. A need for new forms of individuated post-social career and work structures consequently emerged – that is, livelihood systems adapted to the absence of public support and corporate responsibility. Such forces animated the anticipatory 1980s NGO invasion of the global South, which is returned to in subsequent chapters.

Spirit of Capitalism

For some years, it's been unusual to talk about capitalism. Following the high point of anti-capitalist critique at the end of the 1960s, the serious study of capitalism within the academy had ground to a halt by the end of the 1970s (Boltanski & Chiapello 2005: xxxv; Stiegler 2014 [2006]: 11). Through the combined efforts of politicians and academics of all persuasions – from left to right, behaviourists to poststructuralists – the economy became unfashionable. Capitalism disappeared into the environment, becoming an anthropocentric force of nature and as natural as geological change. The timing of this loss of interest was unfortunate as it coincided with a period of rapid and momentous transformation. Capitalism slipped from view during the 1980s and 1990s, just as those far-reaching processes called globalization, deregulation and privatization were fashioning the networked, computer-based and personalized new economy that we now enjoy. This decline in academic interest at such a critical time, to the benefit of nothing except capitalism itself, has made the need to re-engage all the more urgent.

Our view of the new economy has been inspired by Luc Boltanski and Eve Chiapello's important sociological work *The New Spirit of Capitalism* (2005). As the title suggests, they use Max Weber as a point of departure. Most people have to work, whether they want to or not. Because of the burdens and irrationality of this life-long compulsion, capitalism has historically required a set of moral and ethical motivations, beyond money or naked force that justify working. There has been a need, as it were, for a *spirit* of capitalism (Weber 2005 [1930]). For Weber, during capitalism's prehistory, this spirit was originally supplied by post-Reformation Calvinism. Pursuing a secular vocation became a religious duty, and success in the pious creation of worldly wealth was a measure of heavenly predestination. The spirit of capitalism has typically been conceived as an exchange whereby workers, figuratively at least, gave up their freedom for the real or imagined security of work. Struggles over the terms of this exchange were instrumental in shaping labour history. As we shall see, the historic singularity of the new economy is that it reverses this longstanding equation. Workers – or, to be more accurate, agents, operatives, partners and franchisers – are now expected to give up the security that capitalism once provided for the purported freedom of the market.

Writing at the end of the 1990s, Boltanski and Chiapello (2005) identify two modern iterations of the spirit of capitalism and begin to outline the spirit of the new economy then taking shape. Towards the end of the nineteenth century, the spirit had coalesced around the heroic patriarchal figure of the bourgeois factory-owning entrepreneur.[3] Another spirit of capitalism precipitated in the mid twentieth century. This focused on the large industrial company at the centre of the Fordist revolution in standardized mass production and the growth of consumer society. The hero of welfare-Fordism was now the corporate manager, a key figure in the development of standardized mass consumption.[4] Bolstered by a collectivist social state, Fordism provided white managers and key workers with a good deal of relative security in exchange for their freedom. Public education merged with a range of apprenticeship and career openings. Social mobility was a given. There was also a strong company intergenerational bond. Climbing pay grades matched the ageing process and, importantly, in many countries, the mortgage finance cycle. Home ownership joined holiday pay, maternity leave and company pensions to supplement the public support offered by the welfare state (Boltanski & Chiapello 2005: 87). Large Fordist companies also played a significant social role in their own right. Subsidized canteens, family social clubs, sports facilities, holiday camps and pensioner outings were common (Franklin 2016). During the Cold War, when the international plane was still a space of political possibility, welfare-Fordism was a showcase for capitalist security and mass consumption.

Against hierarchy

Welfare-Fordism, however, was also the focus of sustained anti-capitalist critique. Global in its reach, this disparate movement embraced intellectuals, students, young workers, and immigrants. Besides exploitation in the classical sense, the May '68 critique also addressed existential, creative, race and gender issues through its key mobilizing terms of 'alienation', 'hierarchy', 'imperialism' and 'patriarchy'. Through bureaucracy and technocratic managerialism, the one-dimensional corporate world stifled individuality and imagination. Mass consumption was alienating, regimenting and inauthentic. At the same time, this whole rotten edifice was being maintained by the violent suppression of resistance in the Third World (Marcuse 1968). Welfare-Fordism also inherited the work/family divide from earlier iterations of capitalism. Discrimination against women was systemic. Racial segregation was also rife, further privileging the position of white managers and key workers (Duffield 1988). Besides the economic critique of Marxism, the May '68 movement was distinguished by its embrace of women and gay liberation, together with anti-imperial and race equality struggles.

Despite the demonstrations, occupations and flag-waving, the May '68 movement failed to precipitate the revolution. Ironically, however, the critique's sustained condemnation of grey hierarchies, corporate stultification and patriarchy, at the same time as advocating a multicoloured alternative vision of autonomy, authenticity and creativity, played an important role in transforming Fordism into the new economy of the 1990s. With the support of parties traditionally from the left, as Boltanski and Chiapello (2005) document, the energy and aesthetic flair of this critique helped to transform Fordist corporate management structures and labour processes (see also Turner 2006). The onslaught against alienation justified the flattening of hierarchies, extending more autonomy and personal responsibility to key subordinates and projectizing the labour process. At the same time, it eased the way for the incorporation of equal opportunity and anti-discrimination legislation. This transformation, at least in core companies, gave managers and key workers more fulfilment through responsibility, at the same time as beginning to make capitalism more socially inclusive. Importantly, however, it also allowed capital to recoup its near loss of control over shop-floor production during the 1970s (Duffield 1988). In place of dealing with strife, managers were learning the new art of governing through freedom (Rose 2000). The appeal to personal freedom and creativity cut across and dissolved the old left/right divisions of social democracy. In the early 1980s, for example, in France it was the political left that introduced the first liberal market reforms, while in Britain, under the Thatcher government, it was the right. A new centre ground of interchangeable economic caretakers appeared, within which political elites now remain seemingly entrapped.

Progressive neoliberalism

The recoupment of the revolutionary impulse is central to what Nancy Fraser, in the light of the recent electoral victory of Donald Trump, has called 'progressive neoliberalism' (Fraser 2017). She describes this seemingly perverse alignment in contemporary US terms:

> progressive neoliberalism is an alliance of mainstream currents of new social movements (feminism, anti-racism, multiculturalism, and LGBTQ rights), on the one side, and high-end 'symbolic' and service-based business sectors (Wall Street, Silicon Valley, and Hollywood), on the other. In this alliance, progressive forces are effectively joined with the forces of cognitive capitalism, especially financialization. However unwittingly, the former lend their charisma to the latter. Ideals like diversity and empowerment, which could in principle serve different ends, now gloss policies that have devastated manufacturing and what were once middle-class lives. (2017)

While the circumstances of the Trump election and also the UK referendum decision to leave the EU have drawn attention to this alignment, as Fraser alludes, it has wider resonance. Among other things, it goes to the heart of the historical transformation of Fordism to post-Fordism and the formation of the libertarian new economy (Barbrook & Cameron 1995; Turner 2006). While the traditional parties of the left may have failed to do the analytical work necessary to understand the redeployment of capitalism, they have embraced the new economy and liberalization of society (Boltanski & Chiapello 2005: xliii). Progressive neoliberalism is an identitarian stance that typically defines itself *against* 'neoliberalism'. This is usually a shorthand for the assorted ills and excesses of privatization, profit-seeking, welfare cuts and the casualization of work. With its suggestion of a deviation from an acceptable but usually unstated degree of capitalism, such excess is better understood and addressed as what it is: the face of contemporary capitalism itself.

Our conception of neoliberalism draws on the work of Friedrich Hayek (Hayek 1945), which, in the form of behavioural economics, has been in vogue ever since the financial crisis of 2008 cast doubt on the rationality of *Homo economicus* (World Bank 2015: 5; also Cooper 2011; Alcock 2016). Central to Hayekian neoliberalism is the proposition that, due to the complexity of the surrounding world, apart from immediate circumstances, from the standpoint of their individual realities human beings can have little real or comprehensive knowledge of the world around them.[5] This state of necessary ignorance is 'fundamental to neoliberalism' (Spencer 2016a: 17).

Given the contrast between knowledge and data, it can be appreciated that the variously cognitively challenged, attention-deficient, unperceptive or non-reflective subject – or what generically could be called *Homo inscius* – is a citizen of the world of data. Indeed, it is the ideal or necessary citizen. Cybernetics had the foresight, at the moment of its World War II emergence, to have black-boxed human beliefs, intentions and motivations as irrelevant for future prediction (Wiener 1954). More important was the mathematical recording of past behaviour – the more the better – in order to form an objective opinion on intentions. Cybernetics is thus not fazed by neoliberalism's necessarily ignorant subject. By devaluing reason in favour of observable and recordable behaviour, cybernetics has been the primary tool for subjugating, and thus knowing, *Homo inscius* and, as a result, fabricating and shaping its complex emergence (Spencer 2016a). Indeed, cybernetics is the basis, not only of artificial intelligence and complexity-thinking, but of all the cognitive sciences that form the dominant understanding of human behaviour (Dupuy 2000).

Replacing the rational *Homo economicus*, the pivotal player within the new economy, and its aesthetic and pyscho-social governance structures, is *Homo inscius*. Douglas Spencer (2016a), for example, has shown how governing and managing the emergent properties of necessary ignorance are key design inspirations within contemporary architecture. It is also the basis of behavioural economics as well as informing post-humanism. Because of *Homo inscius*' limited ability to perceive, experience or reason, it *demands* the compensatory prosthetic of infrastructural connectivity and *requires* the signals and nudges of an enfolding interactive environment (Stiegler 2016). It is at this point, however, that one detects a tension within progressive neoliberalism. The progressive demand for freedom, authenticity and creativity run counter to Hayek's necessarily ignorant subject. Progressive neoliberalism appears to nurture a demand for fulfilment that outstrips the cognitively challenged abilities of *Homo inscius*. This tension is resolved once it is realized that progressive neoliberalism defines a subjectivity that only exists through personal consumption. It has no existence outside of the continuous demonstration of authenticity through consumption. As such, progressive neoliberalism is entirely lacking in freedom, authenticity or creativity.

Death of the social

Network capitalism is more than simply the spatial restructuring of a Fordist model of capitalism. Central to this restructuring is what Nicolas Rose has called 'the death of the social' (Rose 1996). That is, the abandonment of the long modernist tradition of social reform that interpolates Foucault's analysis of the biopolitics of population (Foucault 1998 [1976]; Foucault 2007). Population in this sense is an abstract statistical construct, not a collection of individual thinking humans. Like a non-sentient biological species, a human population is known through its various registers of normative averages, bell curves and statistical correlations between things like location, birth, education, occupation, circulation and morbidity. The early nineteenth-century discovery of such norms and correlations began a long process of social, urban and political reform in the global North that reached its zenith in the welfare-Fordism of the mid twentieth century (Rabinow 1995).

The death of the social has dramatically changed the spirit of capitalism. As already argued, previously workers gave up their notional freedom for the real or imagined security promised by a globalizing capitalism. Post-Fordism is markedly different; it reverses this longstanding equation. As discussed further in chapter 11 and the Conclusion, the death of the social changes the nature of biopolitics from the normative to the individual and disaggregated register of the biohuman (Evans & Reid 2014). If the species life associated with normative biopolitics could be likened to that of a herding animal, the new biopolitics invokes the characteristics of a predator.[6] Securing the welfare of the biohuman is increasingly concerned with matters of personal lifestyle and maintaining physical and mental fitness. For losers in this regime, at most there are basic safety-nets involving the minimum amounts of personal cash, nutritional intake, energy and medical inputs necessary to maintain clinical life. In place of the social, a resilience regime has been erected on the commons between reality and the world. Chapter 8 examines how network capitalism, in making the economy the site of permanent emergency, has brought the dynamics of the global North and South into alignment with the creation of an expanding global precariat.

The security that Fordism and the welfare state once provided has been exchanged for the promise of a purported freedom beyond the norm (Dean 1999; Rose 2000). In the interests of personal fulfilment, streamlining and authenticity, the spirit of the new economy seeks to free people from all that would deny, slow down or falsify their creative inner essence – not least the standardization, hierarchy and patriarchy associated with the past. With all the appearance of having been designed for social media, this is the spirit of progressive neoliberalism. The emergence of the spirit of personalized freedom coincides with the disappearance of capitalism's social role. As well as spatially restructuring, capitalism 'disappears' in the global North at the same time as companies start to close their subsidized canteens, sell off the sports fields, shut the social clubs, withdraw jobs for life and casualize employment. By the 1990s, network capitalism's signature of inequality, youth poverty and indebtedness was already established (Boltanski & Chiapello 2005: xli–xlii). While places of work still exist, they now provide relatively little security (TUC 2017). Work tends to be a transferrable record of a person's 'passion' for the job and thus an indicator of their potential 'employability' (Cederstrom & Fleming 2012). How individuals trade on this record depends on their creativity, self-promotional skills and resilience in the new post-social wild.

Exchanging security for the personal freedom of the wild is no small thing. It should, on the face of it, have been a hard sell. That it wasn't should not disguise that it is a weak and fractious spirit of capitalism. It has forced individuals to shoulder the transaction costs that companies and public bureaucracies once absorbed – a burden transfer that has been streamlined with the increasing automation of daily life (Stiegler 2016). The new spirit promises the future to those who are willing to become entrepreneurs of the self. This spirit no longer focuses on the world of work – indeed, it no longer focuses on production; its driving concern is consumption, unmediated experience and the right to a lifestyle of choice. The new hero is not the Fordist company manager of the past; it's the scalable figure of the entrepreneur as a networker and celebrity (Boltanski & Chiapello 2005: 357–60) – that is, a person able to bend to their advantage the spatial displacement and asymmetries of information and power inherent within the connectivity of network structures. Their reward is personalized consumption: the ability to choose the when, where and how of living through consuming – even moving to New Zealand should things start to fall apart elsewhere (Donnell 2017).

Mobility v. immobility

It has already been argued that circulation and connectivity are different yet interconnect and affect each other. An important real-world effect of growing connectivity, for example, has been an inexorable increase in the speed of circulation (Virilio 2007 [1977]). As if trying to catch up with connectivity, people and things move faster over greater distances, markets work 24/7, and human bodies adapt to hyper-urbanized milieus that never sleep (Crary 2014). The need for speed associated with connectivity has outmoded and transformed earlier social distinctions based on class (Boltanski & Chiapello 2005: 360–5). While having residual effects, class is realized today more through the distinction between mobility and immobility; the fast and slow; the free to move and those who are territorially entangled. Differences and combinations of education, wealth, age, race, gender and location now resonate with and are operationalized through the connectivity registers of speed and mobility. These differences are amplified by the changing nature of international space. Now corrugated by migration barriers and anxiety checks, the international has separated into slow and fast lanes (Petti 2008). *Sans-papiers* are hampered and contained at borders and checkpoints. Although slowed and differentiated by security advisories, insurance premiums and wealth, *avec-papiers* are still able to move along designated global circuits within approved and monitored carriers. Despite the growing stress of air travel, having once got there, connectivity enables the luxury of being able to leap across the intervening ground friction and no-go areas through a network of screen interfaces. As international space has become increasingly corrugated, multilevelled and striated into speed differentials, the most compelling spatial metaphor for envisioning this layered apparatus requires a leap to the 'vertical' plane (Weizman 2002; Elden 2013). Increasing ground friction has anticipated the verticalization of political geography.

With capitalism slipping off the academic agenda, the notion of exploitation, like class, has also been problematized. In the past, the existence of exploitation was taken for granted. The public aim of social democracy was to manage the redistribution of wealth between social classes. There were government tribunals, wage councils and public watchdogs to monitor this process. During the 1980s, the notion of exploitation became unfashionable and was replaced by 'social exclusion' (Boltanski & Chiapello 2005: 354–5). Although exclusion has been the focus of much debate and public concern, exclusion is not exploitation. With high levels of under-employment and job insecurity, many of the excluded are, effectively, out of regular work. Viewed conventionally, you have to be in work to be exploited. The spirit of progressive neoliberalism is meritocratic. You get back what you put in, it's only fair. In Britain, for example, the distinction between 'strivers' and 'skivers' has bitten deep into popular and political culture (Jones 2011). The skiver is no longer a figure requiring social justice. Instead, it's a question of the just deserts of a meritocratic individual justice. As a concept, exploitation was more progressive. It does not blame the exploited but recognizes the necessity of their position. More than sympathy, solidarity is demanded. For the excluded, however, while their plight is a cause for concern, within a meritocracy they cannot escape some burden of personal responsibility for their predicament.

Boltanksi and Chiapello (2005) have attempted to rejuvenate the idea of exploitation[7] in relation to network connectivity. Between the different registers of mobility and immobility, power, influence and profit accrue to those networkers able to exploit the spatial displacements and asymmetries of information inherent within network morphologies. This is what they call the 'mobility differential' (2005: 371). Chapter 6 examines this idea in relation to NGOs and developmentalism. The spirit of progressive neoliberalism's celebration of speed, mobility and connectedness operates to devalue and depreciate those skills, conditions or experiences that are territorially entangled, locally based or otherwise immobile. Mobile, interchangeable and comparative cosmopolitan skills are valued more than fixed, grounded or left-behind rust-belt expertise. Core companies, always on the point of threatening to leave, use their mobility differential to wring financial concessions and tax breaks from governments, at the same time as they impose regressive conditions on territorially entangled sub-contractors and workers. Despite this situation, however, the new spirit holds that the future security and wellbeing of the immobile and excluded lie in the creativity, connectivity and philanthropy of the hyper-mobile (Zuckerberg 2014). Materially, however, the dependency is the other way around. Even if dispersed across globally scattered sites of multiple and minor monadic transactions, without the devalued contributions of the little people, the big people would not be who or what they are.

For the little people, the slow and left-behind, network capitalism's capture of the mobility differential has translated into decades of declining wages, falling living standards and the growing casualization and abjection of work within the global South *and* the North (LeBaron & Ayers 2013; Corlett 2017). Not only are mobile private and corporate actors able to impose regressive conditions on the immobile, as the 2008 financial crisis demonstrated (UN 2011), they can also pass on to the territorially entangled the brute force of adverse conditions, shortages or market downturns. In other words, the mobility differential renders the precariat vulnerable to the direct and unmediated uncertainties of their surrounding environment. The death of the social in the global North has blurred with its absence in the South to render such uncertainty a general biohuman condition. Capitalism slips from view at the same time as it becomes a force of nature. It is now just as uncontrollable, unpredictable and, in a sense, unremarkable as the weather. Ideas of 'risk', 'complexity' and 'uncertainty' have seamlessly colonized our view of the outside world (Beck 1992 [1986]), displacing earlier modernist notions of certainty, predictability and protection (Hewitt 1983). Yet, rather than uncovering some superior or essential ontological essence, what we are experiencing is the permanent emergency of capitalism itself. The post-humanist indistinction between reality and the world, and its insistence on an unmediated relationship between a subject and its enfolding environment, finds its sobering real-world condition in the growing phenomenon of precarity.

Conclusion

This chapter has been concerned to draw out the action-oriented will or spirit behind the transition from welfare-Fordism to the new economy. While security in the latter may have been sacrificed, the intention has been to illustrate the progressive drive, the will to freedom and authenticity, that not only tore down the professional hierarchies of Fordism but still resonate in the identitarian politics of today. This spirit is returned to when the boomerang effect interconnecting the global North and South is examined in chapters 5 and 6. The next couple of chapters, however, focus on the resistance to capitalism. Rather than just seeing the May '68 critique as recouped in progressive neoliberalism, they explore what may have been lost in that incorporation. The May '68 movement had its own critique of modernism. This tends to get buried in the triumphalism of a conservative post-humanism (Latour 2004). The following chapter examines the overlaps and differences between cybernetics and the Marxism of the New Left in confronting the loss of faith in the sovereignty of the subject. In the form of *Homo inscius*, this loss is fundamental to the new economy.

Chapter 3

ENTROPIC BARBARISM

The term 'New Left' gained currency in the West at the beginning of the 1960s (Mills 1960). Rather than representing a coherent movement or discrete set of institutions, the New Left is best understood as a historic and contingent political impulse that drew together a loose and dynamic coalition of socialist, Marxist and countercultural elements that together helped shape the liberal race, class and sexual revolutions that define the present. Besides opposition to the cultural status quo, these different tendencies shared a distrust of the incorporated and bureaucratic old left, including the institutions of organized labour, the Western communist parties and the Soviet Union. The New Left was an important contributor to what was called the May '68 movement or critique in this book's Introduction. While having lost what coherence it had by the mid-1970s, the irony of the May '68 movement's contribution to the transition from Fordism to post-Fordism has already been noted. The US counterculture, itself an early fellow-traveller with the New Left, is also implicated here. In particular, it transformed concerns that computers were a new means of corporate control and governance into the embrace and celebration of digital technology (Turner 2006).

While this corporate recoupment is important, a consideration of the New Left in its own right is still warranted. Between the mid-1960s and early 1970s, it was active in developing a critical understanding of what the Marxist philosopher Herbert Marcuse had called 'technological society' (Marcuse 1968). Prescient in relation to current concerns regarding the growing hegemony of post-humanism, data behaviourism and algorithmic governmentality (Rouvroy 2012;

Halpern 2014a; Stiegler 2016), today such opposition is all but forgotten. This is especially notable given the fact that, for a number of years, the May '68 critique had a raucous and recalcitrant presence on university campuses in Europe, the USA and Japan (*NLR* 1968). Of particular interest to us is that the New Left's view of revolution was predicated on circulation rather than connectivity – or, at least, a connectivity that was subordinate to circulation and ground truth. This critique belongs to a world that is now irretrievably gone. Recreating some characteristics of this world, however, is important because they indicate what has been lost in the rise to dominance of a cybernetic *episteme* and the consolidation of network capitalism.

This chapter examines the autonomy and openness to the circulation of political practice that furnished the New Left's transitory conditions of existence. It then considers how both the New Left and cybernetics were part of the same post-World War II questioning of the sovereignty of the subject. Whereas the New Left, through the negative dialectic, saw the possibility of capitalism's descent into barbarism, for cybernetics the challenge was to avoid entropy. While originally overlapping in terms of their social concerns, it was cybernetics that triumphed, thus helping to create the conditions for network capitalism.

Precursors of Revolution

After decades of post-colonial anxiety and inhibition regarding the outside world (Sloterdijk 2013 [2005]), the conventional wisdom that the international arena is now a space of danger, volatility and unknowable risks has been effectively normalized among the liberal intelligentsia (Goodhart 2017). To preface a policy brief on the future of global security, for example, without some incantation regarding 'uncertainty' or 'complexity' would risk appearing out of touch. So complete is *Homo inscius*' disenchantment with the world, it is worth reminding the reader that, even within living memory, this was not always so. From the end of World War II until the mid-1970s, for progressive opinion at least, decolonization was a time when the international was a positive space of political opportunity and revolutionary hope. With the historical agency of the metropolitan working class now dissipated, the possibility of radical renewal – indeed, world revolution – was vested in the colonies and former colonies.

During the course of the 1950s, the many socialist-oriented national liberation and anti-colonial struggles had inspired the emergence of the 'Third World' as a progressive geopolitical entity (Woodis 1960; Buchanan 1963). Within a couple of decades, however, this term, as well as the emancipatory hopes and revolutionary imaginary associated with it, had effectively disappeared (Duffield 2001: 22–5).

Geographically, the Third World was diverse, ranging from tropical forests and savannah regions to coastal littorals, mountains and deserts. Home to around 2 billion people, then two-thirds of the world's population, still mostly living in the countryside, it was made up of a mix of independent countries, semi-colonies and colonies at different levels of infrastructural development, with varying political histories and cultural backgrounds. For the New Left, cutting across this diversity was a common legacy of past humiliation inflicted by colonial exploitation, a legacy that bound these countries together 'in a vast "fellowship of the dispossessed", a "commonwealth of poverty"' (Buchanan 1963: 6). Moreover, for most of these countries, independence had not been freely granted. It was won through struggle and violent confrontation with colonial powers often reluctant to relinquish control. Given the agrarian character of many of the liberation movements involved, it became axiomatic for the New Left that, in the changed conditions of the mid twentieth century, the peasantry had become 'the truly revolutionary class in colonial countries' (1963: 11). At a time when political violence could still be justified within the academy as a means of promoting progressive social change, the peasantry was praised for its ability, when faced with implacable opponents, to discover quickly 'that violence, and violence alone, pays off' (1963: 11). Having no seat at the table, and frequently despised by the educated classes of the town, the peasantry was regarded as having nothing to lose from the sundering of the colonial and neocolonial chains that bound them. Such enthusiasm, however, could also encourage, as Herbert Marcuse cautioned against, an uncritical left romanticism that regarded 'national liberation movements as the principal, if not the sole revolutionary force' (Marcuse 1969: 29).[1]

During the 1950s, in a diverse mix of countries, including China, Vietnam, Cuba, Kenya, Algeria, Brazil, Angola and Egypt, the peasantry had emerged as a driving political force in the struggle for national liberation. Import substitution, land redistribution and policies for national food self-sufficiency were the order of the day

(Bush 2007). Importantly, for both the left and right of the political spectrum, the expectation was that this push for social liberation would, during the course of the 1960s, continue to spread in Latin America, the Caribbean, Africa and Southern Asia (Marcuse 1969: 22). Given intellectual coherence by the Truman Doctrine, the USA and its allies launched a range of counterinsurgency interventions that, besides economic embargoes and aid programmes, often involved the covert support of violent military proxies (Bostdorff 2008).[2] The aim was to contain and neutralize these struggles, shaping them to Western interests. While the number of international armed conflicts under way in the mid-1950s numbered around fifteen, most of which were externally aggravated civil wars, the incidence of such conflicts would rise steadily until peaking at around fifty by the end of the Cold War (PRIO 2009). By the beginning of the 1970s, however, as decolonization began to draw to a close, the survival of the revolutionary impulse within the Third World was already seen as tenuous – and, with it, the chance of opposing the growing domination of technological society (Marcuse 1972).

The New Left's foregrounding of the world as a space of political possibility has proven to be a historically contingent proposition. An autonomous peasantry, for example, was a transient phenomenon. During the 1950s, much of the Third World was still largely agrarian. By the 1970s, however, this was changing fast (Hobsbawm 1994). As a result of the expansion of commercial agriculture, land expropriation, conflict and the lifting of colonial restrictions on urbanization, an informalized and precarious world of slums and shanty towns, integrated into the circuits of capitalist dependency, would soon displace ideas of an independent peasantry (Davis 2006). Indeed, by the 1980s, like the 'Third World', the term 'peasantry', once ubiquitous in area studies, had also disappeared, to be replaced, as we shall see, by contemporary ideas of poverty and 'the poor'.

In comparison with the present, the period of decolonization was also a time of relative openness to the international circulation of political praxis. For example, anti-colonial networks and national liberation solidarity campaigns were common. Western European countries had been moved to accept permanent immigrants from their colonies and former colonies, together with allowing refugee settlement and recruiting significant numbers of migrant workers. For a while, aspirational white settler colonies such as Australia, New Zealand and Canada also lifted the 'colour line' that had earlier applied, especially towards Asian labour migrants (Meyers 2002). Immigration and refugee controls in Europe were only beginning to crystallize in the mid-1960s.[3] During the following decade, however, this would rapidly turn towards a programme of restriction, closure and return. Together with an autonomous peasantry, this transitory circulatory window was also a part of the New Left's contingent existence.

Reflecting the anti-colonial ethos among the intelligentsia – unlike today's trend of risk aversion and remoteness (Collinson & Duffield 2013) – academics and students from Europe and the USA were a common feature within the universities of newly independent countries.[4] Either working full-time, on sabbatical leave or as researchers, most of these academics emulated the progressive Third Worldist ethos of these institutions. Although soon to descend, the cultural curtain that existentially divides what are now the global North and South was still open. It left a window for the flow of radical ideas and political solidarities, as opposed to aid commitments, emotions and ethical postures (Chouliaraki 2013). Academics, however, were not the only ones caught up in this anti-establishment *Zeitgeist*. As the Vietnam War attests, Western journalists had yet to embed within systems of military accountability (Page 1989). Still in their infancy, satellite TV and video technologies were unencumbered by security restrictions. An independent and grounded media brought the raw and unguarded images of the Vietnam War into people's homes on a nightly basis. This relative openness and the mode of shared connectivity allowed the political effects of Third World liberation struggles to ripple across the globe. Through anti-war and solidarity campaigns, these ripples interconnected and energized progressive opinion in Europe and the USA – not least among students who, with the expansion of a grant-assisted and debt-free system of higher education at a time of full employment, had the autonomy necessary to develop critical thought and independent action.

The Sovereign Subject

While the ending of World War II would unleash a modernist will to rebuild and reconstruct anew, among many it also fostered a sense of pessimism and failure. It was clear, for example, that liberalism, for a long time a beacon of Western civilization, had failed to prevent fascism and the horrors it unleashed. The leap in technological advancement necessary for victory did not assuage such anxieties. Indeed, the atomic bomb raised doubts about the moral compass that political elites were now using. When he was interviewed in 1971, such anxieties were the beginning of what Foucault called the 'great questioning of the sovereignty of the subject we are witnessing these days in our culture' (Foucault 2013: 43). While this 'great questioning' was shaping cybernetics and neoliberalism, it also inflected the New Left and its break with what C. Wright Mills called 'Victorian Marxism' (Mills 1960). Emerging from the questioning and anxieties of the time, the positions adopted by cybernetics and the New Left were different yet strangely resonant. Should political opposition fail, the New Left saw the danger in terms of capitalism's self-directed descent into 'barbarism'. For cybernetics, leaving society to itself, as with any other circulatory system, posed the risk of increasing 'entropy' or indistinction. The New Left position is examined first.

Negative dialectic

The philosophy of Herbert Marcuse is remarkably prescient with regard to contemporary concerns regarding the apolitical and atheoretical effects of post-humanism and the computational turn (Galloway 2013). Technological capitalism was now able both to create new needs and, importantly, to satisfy them as never before. The ability to guarantee the young a better life than their parents had been important for the incorporation of the labour movement. The price being paid, however, was significant. At the height of the May '68 agitation, Marcuse felt that capitalism was already 'disappearing', in the sense of a capitalism able to exist with an autonomous social realm distinct and outside of itself. Technological society was synonymous with the increasing primacy of logistical reasoning and the devaluation of the individual. Rights and liberties were 'losing their rational content' to the demands of a one-dimensional technological rationality (Marcuse 1968: 19). Independence of thought, autonomy, the right to political opposition and an independent critical function were being lost to the pyscho-social governmentalities made possible by mass consumption. By dint of this capacity, the emerging capitalism 'may justly demand acceptance of its principles and institutions, and reduce the opposition to the discussion and promotion of alternative policies within the status quo' (1968: 19). While Francis Fukuyama (Fukuyama 2002 [1989]) and Bruno Latour (Latour 2008) would eventually celebrate the arrival of just such a society, for Marcuse it would have all the necessary ingredients for the 'the mass basis of a neo-fascist regime' (Marcuse 1969: 31). Unless counter-forces could prevent it, one would see an explosion of internal contradictions 'which would make a re-examination of the concept of revolution a merely abstract and speculative undertaking' (1969: 33).

This danger of a descent into a designer 'neo-fascism' where political hope had died was a consequence of the *negative dialectic*. Associated with Theodor Adorno and the Frankfurt School (Buck-Morss 1977), its central motif is that, by the early twentieth century, capitalism had exhausted any emancipatory potential it may once have had, becoming instead the engine of the Enlightenment's self-destruction (Horkheimer & Adorno 1979). For Marcuse, the negative dialectic involved critically addressing the rationalist, even positivistic, strain in Marx, 'namely, his belief in the inexorable necessity of a transition to a "higher stage of human development"' (Marcuse 1969: 33). Although Marx was fully aware of real and possible failures, 'the alternative "socialism or barbarism" was not integral to his concept of revolution' (1969: 33). The negative dialectic, Marcuse believed, must become part of the contemporary revolutionary outlook.[5] Failure is not simply a lost opportunity. It carries the fateful responsibility of having allowed a deepening political, cultural and spiritual barbarism.[6]

The negative dialectic means that those wanting a better world can no longer draw spiritual comfort from Hegelian sentiments that humanity will, eventually, fulfil its destiny and society attain a higher moral stage. With autonomy, knowledge, politics and critique being extinguished along the way, left to itself, capitalism will destroy the Earth and everything on it. For Marcuse, barbarism is not limited to acts of violence, cruelty or ignorance. It is a technological barbarism intrinsic to the increasing automation of capitalism and growing dominance of technoscience within society: 'the subordination of man to the instruments of his labour, to the total, overwhelming apparatus of production and destruction, has reached the point of an all but incontrollable power: objectified, [reified] behind the technological veil, and behind the mobilized national interest, *this power seems to be self-propelling, and to carry the indoctrinated and integrated people along*' (emphasis added, Marcuse 1969: 33).

As if struggling against the incoming sea, the negative dialectic gives a sense of urgency to Marcuse's opposition to corporate capitalism. While Norbert Wiener, the founder of cybernetics, would not have been unsympathetic to such New Left anxieties, as discussed further in chapter 5, it would be cybernetics that would ride the waves.

Resisting entropy

Cybernetics also has its origins in the post-World War II pessimism regarding the human condition. Given the ideological excess of war, from a humanist perspective, cybernetics aspired to furnish a rational mathematical alternative to the irrationality of history. It offered a New World technological solution to the ideological failure of the Old World (Halpern 2014b: 225). In black-boxing human subjectivity – that is, demoting beliefs, intentions and causal reasoning in relation to understanding human behaviour – cybernetics provided an engineered and operational alternative to ideology. Rather than conscious motives or intentions, what was required was a record of past behaviour – the more comprehensive, the better. From this data, the probability of future actions could be mathematically inferred. Cybernetics promised to tame the political unreliability and arbitrariness of subjective beliefs through the objective analysis and management of information. Rather than the barbarism of the negative dialectic, in the work of Norbert Wiener, post-war pessimism expressed itself metaphorically in the thermodynamic concept of entropy.

The new physics to which Wiener subscribed was no longer concerned with what will happen but, given the data and calculative power available in the mid twentieth century, what will probably happen. This probabilistic physics introduced an 'element of chance in the texture of the universe' and was close to the St Augustinian tradition of Christianity (Wiener 1954: 10). The element of randomness and incompleteness now allowed within physics 'is one which ... we may consider evil' (1954: 10). This was not the positive evil of the secretive, devious and calculating demons of the Manicheans, it was the negative or environmental evil of entropy and its associations with randomness, unpredictability and uncertainty. Entropy tends to increase within all closed circulatory systems. Over the long term, systems eventually deteriorate and lose their distinctiveness. They move from the least to the most probable state, and 'from a state of organisation and differentiation in which distinctions and forms exist, to a state of chaos and sameness' (1954: 12). Having a resonance with contemporary resilience-thinking, for cybernetics, entropy is a measure of a system's disorganization; the more indistinction and sameness, the greater the entropy.

Working against the long-term tendency for the universe to homogenize, local enclaves of connectivity exist where, on a limited and temporary basis, countervailing forces resisting the entropy of circulation are encountered. The feedback of information between human and nonhuman milieus 'strives to hold back nature's tendency toward disorder by adjusting its parts to various purposive ends' (1954: 27). While the universe will eventually become a vast undifferentiated temperature equilibrium where nothing happens, there are fleeting but nonetheless important enclaves where entropy is being resisted through an increase in human and nonhuman connectivity and inter-organization. It is with this possibility of resisting disorganization 'that the new science of Cybernetics began its development' (1954: 27). The physical and environmental interactions between individuals and the communication technology then available were, according to Wiener, analogous in their attempts 'to control entropy through feedback' (1954: 28). Cybernetics casts connectivity, or the free and purposive exchange of information between human and nonhuman systems, as the principal means of resisting the entropy of indistinction that circulation, left to itself, will inevitably produce.[7]

As a governmental ontology, first-order cybernetics is a form of *anti*-catastrophism. That is, while recognizing that systems have an inbuilt tendency towards dissipation, rather than welcoming this tendency, as resilience and catastrophism will eventually do (Homer-Dixon 2007), in its initial formulation at least, cybernetics opposed this tendency.[8] In first-order cybernetics, there is a certain moral overlap with the New Left. As Fred Turner points out, embedded in Wiener's theory of society as an information system 'was a deep longing for and even a model of an egalitarian, democratic social order' (Turner 2006: 24). However, the cybernetic approach to entropy was effectively to eliminate the autonomous rational or sovereign subject, thus paving the way for an eventual liaison with neoliberalism's *Homo inscius* (Cooper 2011). The New Left, however, addressed the prospect of barbarism politically by working for an international anti-capitalist alliance based on an appeal to the rational subject. This important difference is explored in the next chapter in relation to structural anthropology. True to the paradox of connectivity, it is sufficient to say here that, while an entropic barbarism now seems closer than ever, and pressing on all sides, the rise to dominance of a cybernetic *episteme* has been comprehensive, seamless and extraordinarily successful. Indeed, it has evolved into an algorithmic 'mode of governmentality without negativity' (Rouvroy 2012: 13) – that is, a mode of governmentality that no longer has to acknowledge that anything of significance exists outside of itself.

As a metascience, cybernetics is a *sui generis* mode of knowledge that uses methods of analogical transfer to discover equivalent modes of functioning in otherwise different orders of reality. It is a totalizing constitutive 'technological mentality' that functions equally across all matter and substances, whether alive or dead, human or nonhuman (Simondon 2009). The rhetoric of first-order cybernetics actively facilitated scientific networking and entrepreneurship. This enabled it to spread across the military-industrial complex and, as resilience would later replicate, it become a 'universal discipline' (Turner 2006: 25). Cybernetics would leap across disciplinary boundaries, creating new personal and institutional networks (Hayles 1999: 50–75): 'If biological principles were at work in machines, then why shouldn't a physiologist contribute to the work on computers? If "information" was the lifeblood of automatons, human beings and societies alike, why shouldn't a mechanical engineer become a social critic?' (Turner 2006: 25).

While the disciplinary domains of biology, psychology, sociology or anthropology, for example, are not usually considered to be machines ontologically, they become so analogically for the convenience of mathematics and their cybernetic reconstitution as technoscientific domains (Halpern 2014a). Cybernetics has authored totalizing constitutive leaps from the emergent and self-ordering tendencies observed in the physical and natural worlds to the human realms of economy, society and the mind, now reconstituted as information processing systems (Hayles 1999; Cooper 2011; Rouvroy 2012). However, since these orders of reality and disciplinary domains are different, understanding them through computational analogy adds little to either grasping their intrinsic differences or conceiving them in terms of inactivity or stasis (Simondon 2009). The technological mentality understands them ecologically, or how they intercommunicate, and, importantly, through the dynamism of their interconnected complex emergence and becoming (Dillon 2007).

That is, as so many functionally equivalent orders of cybernetic immanence and possibility.

Conclusion

That capitalism had lost its progressive role in world history was an established theme by the mid twentieth century. The New Left proposed the revolutionary overthrow of technological society. Cybernetics, however, saw the issue in terms of entropy and the development of information-based command and control technologies. Before considering the rise of the cybernetic *episteme*, the next chapter looks at the academic resistance of the late 1960s and early 1970s to the encroaching empiricism and behaviourism associated with it. Moreover, by way of demonstrating the relative openness to international circulation that existed at the time, it outlines the types of structural analysis and immersive, critical anthropology that were then still possible.

Chapter 4

BEING THERE

While the universities have been instrumental in the rise and dissemination of the cybernetic *episteme*, during the 1960s, compared to the economy, they were still not fully incorporated within the circuits of capitalism.[1] This, and other factors, made the academy a strategic 'weak link' in the revolutionary struggle. Today, it is widely argued that, over the coming decades, automation and the growing sophistication of machine learning or artificial intelligence will accelerate the elimination of swathes of erstwhile middle-class professional and knowledge-based service employment (Brynjolfsson & McAfee 2011). The beginning of the 1960s can be seen as the opening or expansionary phase of this destructive cycle. Basically, it was the time when those professional and knowledge-based jobs and services that have already disappeared, or are now under threat, were fast expanding. Structured around welfare-Fordism and Keynesian full employment, from the USA to Europe and beyond, there was a new and growing demand for university-educated professionals, technically qualified administrators and knowledge brokers necessary to create and manage a globally expanding consumer economy in all its varied practical and cultural modalities.

This chapter examines the role of the intelligentsia as the New Left's 'lever of history', and the forms of struggle waged within the academy against an encroaching empiricism and behaviourism. Important here is the rejuvenating effect that the prospect of world revolution had upon the colonial discipline of anthropology. It concludes with the review of the immersive fieldwork and structural methodology that were still possible in the mid-1970s. This world of history, causation and political solidarity would be rapidly swept away during the deepening crisis of modernism. The NGO-led invasion of the global South the following decade was an important harbinger of the world to come.

Theoretical Practice

Drawing on the failures of the labour movement to effectively oppose World War I or prevent the rise of Nazism in Germany, the New Left's political view of the metropolitan working class was largely dismissive. While a place still existed for 'unincorporated' workers, institutionally the working class was no longer the 'necessary lever' of world revolution (Mills 1960: 22).[2] The transfer of this responsibility to the Third World was described in the previous chapter. In terms of replacing the agency of the working class, perhaps reflecting a premonition of its coming extinction at the hands of *Homo inscius*, this role fell to the intelligentsia. In 1960, C. Wright Mills was already arguing that the youth, including middle-class students and intellectuals, were emerging as agents of cultural and political change. In Turkey, South Korea, Cuba, Taiwan, Okinawa, Japan, and – as members of CND – in Britain, students were making themselves heard. The same was true of the Soviet bloc. Students, academics and writers were raising their voices in Poland, Hungary and Russia. In reflecting on the problem of agency, 'I have been studying, for several years now, the cultural apparatus, the intellectuals – as a possible, immediate, radical agency of change. For a long time, I was not much happier with this idea than were many of you; but it turns out now, in the spring of 1960, that it may be a very relevant idea indeed' (Mills 1960: 22).

Given the rise of the US counterculture, and its visible rejection of the straight world of mass consumerism, this view of the strategic importance of the intelligentsia would grow. The youth opposition in the USA could have a political effect because it was:

> free from ideology or permeated with a deep distrust of all ideology (including socialist ideology); it is sexual, moral, intellectual and political rebellion all in one. In this sense it is total, directed against the system as a whole: it is disgust at the 'affluent society', it is the vital need to break the rules of a deceitful and bloody game – to stop co-operating any more. (Marcuse 1967: 7)

The internal opposition to corporate capitalism was no longer organized labour but 'the two opposite poles of the society' – namely, in the ghettos and among the middle-class intelligentsia, especially students (Marcuse 1969: 30). While, in themselves, these actors could not initiate radical change – needing alliances with unincorporated workers and international revolutionary forces – for the New Left, the intelligentsia was the new agent of history.

During the 1960s, the universities were a contested space in the rise of the cybernetic *episteme*, or, as Marcuse called it, an emerging 'technological rationality' (Marcuse 1968: 25). Central to this rationality was the early post-human blurring of a distinction between an inner reality and an outer world. The inner was being transformed into the outer, with the consequent loss of an internal private space, under the techno-psychological weight of corporate capitalism's claim on the entire individual. Industrial psychology, once confined to the factory, was increasingly being used to get inside the heads of once autonomous customers (Packard 1957). The loss of an inner private space, 'in which power of negative thinking – the critical power of Reason – is at home, is the ideological counterpart to the very material processes in which advanced industrial society silences and reconciles the opposition' (Marcuse 1968: 25).

In anticipation of the celebratory arrival of post-humanism, this technological rationality could be detected in the trend towards operationality in the physical sciences and behaviourism in the social sciences. Common to both 'is a total empiricism in the treatment of concepts; their meaning is restricted to the representation of particular operations and behaviour' (1968: 27). This operational tendency within philosophy, psychology and sociology was the means by which critical concepts were being eliminated through claims that no adequate account of them could be given in terms of their operations or behaviour. Preparing the way for *Homo inscius,* what Marcuse called a 'radical empiricist onslaught' within the universities was providing the methodological justification 'for the debunking of the mind by intellectuals – a positivism which, in its denial of the transcending elements of Reason, forms the academic counterpart of a socially required behaviour' (1968: 28).

Throughout the 1960s, as the ghettos burnt, student campus unrest grew. The unexpected eruption of the May '68 events in France, moreover, confirmed the centrality of students and the university campus as an autonomous site for revolutionary agitation.

For politicized students, their relative openness provided an intellectual refuge against a grey and alienating corporate world. At the same time, however, tensions were growing between the espoused independence and intellectual freedom of the academy, as a public space where contrary artistic or philosophical temperaments could stand outside and contest the norm, and what was the beginning of its subordination to the skill and vocational requirements of technological capitalism. This was captured in the widespread caricature of the 'Fordist University' as a conveyor belt endlessly turning eager and idealistic students into regimented lines of zombies holding degree certificates.[3] At the same time, while the valuable societal role of students was being extolled, universities were administered in an autocratic and paternalistic manner, alongside 'the contradictions within bourgeois ideology itself which make sociology faculties paradoxically the producers of Marxist revolutionaries, and so on' (NLR 1968: 5).

Writing in 1968, Anthony Barnett could, in all seriousness, advocate that UK university campuses could be transformed into democratic 'red bases' operating across faculties, departments, halls of residence, flats, university societies and newspapers. Following the model of national liberation in the Third World, these bases would establish 'the physical liberation of student existence from external controls' (Barnett 1968: 44). Using this autonomy, students would be able to reach out to other radical and revolutionary forces. This view emerged on the back of a wave of student occupations. In the UK, the ostensible reasons for such actions were often relatively minor – for example, administrative restrictions, accommodation policy or procedural matters that were felt to infantilize students. This agitation, however, was also international in scope. At the time of the Tet offensive, when the Vietnamese peasant was breaking the back of the US military, the European student movement was seen as having joined 'the students of Asia, Africa and Latin America in the revolutionary struggle. There was a time when the occupation of universities in Japan, Iran or Argentina seemed part of a separate realm of politics, divorced from the realities of Europe. Now Berlin, Rome, New York, London, San Francisco stand alongside Tunis, Montevideo and Mexico City' (NLR 1968: 5).

The struggle, however, was also being waged ideologically against the 'empiricist onslaught' of behaviourism in the social sciences

(Blackburn 1973). As reflecting 'bourgeois sociology', their ideological content was:

> painfully obvious. They refuse to admit that the mind can act except according to the most shallow and stunted bourgeois 'common sense' and 'rationality'. All must seek equilibrium and order. Every possibility of disorder and unreason must be discounted and expelled. Danger must be suppressed or corrected by homoeostatic mechanisms. It is not difficult to understand why these theories flourish in the universities. (NLR 1968: 2)

The political praxis being developed through such critique was distinguished from abstract speculation. Theory was only as good as its ability to mobilize people and translate ideas into practice. To the extent that it could overcome ideological noise and become a force of resistance and change, theory constituted itself as power (Marcuse 1969: 28). The possibility of theoretical practice, of constantly challenging the empiricism of the day, made the academy during the late 1960s and early 1970s – especially sociology departments – a bruising political battleground. Wholly dependent on the historic window of opportunity existing at the time, in retrospect it was the last concerted attempt by the intelligentsia to resist the growing dominance of capitalism over society and the penetration of a one-dimensional technological rationality into previously autonomous areas of social life.

New Anthropology

For the New Left, the international was a circulatory political whole; the Marxist theoretical framework was global in scope. If Vietnam was part of the international system of corporate capitalism, then national liberation in the Third World became integral to the wider socialist revolution. Importantly, however, the struggle for national liberation was creating a political alternative to both Western capitalism and Eastern socialism, especially that of the Soviet Union. The hope was for a different way of socialist construction not just 'from below', 'but from a "new below" not integrated into the value system of the old societies – a socialism of co-operation and solidarity, where men and women determine collectively their needs and goals, their priorities, and the method and pace of "modernization"' (Marcuse 1969: 29–30).

Central to the New Left position was that the peasantry had still to be fully incorporated within capitalism. It was because they were outside, as it were, and still 'the human basis of the social process of production' (1969: 31), that they could furnish an alternative model of socialism. With the world a circulatory whole, achieving this alternative would require synchronization between the negating external forces and those internal forces seeking change. Moreover, this synchronization could never be 'the result of organization alone, it must have its objective basis in the economic and political process of corporate capitalism' (1969: 31). In other words, it would require an understanding of how global capitalism, now given a new lease of life by decolonization, was spreading in the Third World. Among students of anthropology, this proposition had a radicalizing effect.

Energized by the May '68 critique, radical students and academics had declared anthropology to be in terminal crisis (Banaji 1970). Still to confront its colonial heritage, and still trapped in the study of so-called primitive societies, its theoretical base had stagnated. The ahistoric empiricism of functionalist anthropology had long ignored the asymmetries of international power that constituted the conditions of its existence (Asad 1973). While a few anthropologists proved uncooperative, colonial anthropology had developed as an applied discipline geared to refining domination. The studying of the social institutions of the 'subject races', for example, informed the creation of systems of native administration (Macmichael 1934; Lugard 1965; Johnson 1982).[4] As an attempt to govern through 'traditional' political structures, even creating them when they did not properly exist, native administration was liberal imperialism's main response to the growing nationalist demand for the modernity that came with independence.

Following the onset of decolonization, the conditions that had supported colonial anthropology had all but disintegrated by the beginning of the 1970s. The political emergence of the Third World had, by force of circumstance, 'thrust itself obtrusively into the anthropologists framework' (Asad 1973: 13). With its disruptive view that the peasantry was now the agent of history, the New Left critique, by 'transcending itself' (1973: 18), made it possible for anthropology to throw off its colonial heritage and become something radically different. Rather than an outsider helping to refine reactionary systems of governance, the new anthropologist was now a sympathetic ally of the exploited against such control. Radical anthropologists not only recorded and participated in the struggles of the downtrodden, they would also attack 'the role of anthropology in creating, preserving and implementing ideologies of oppression' (Faris 1973: 170). Functionalism was not only rejected because of its empiricism, it was dismissed in solidarity because 'the people on whom this sort of theory was applied are rising up to change it' (1973: 166). The new anthropology would help bring about progressive social change.

Radical students of the international, for example, created networks and journals to support national liberation and oppose Western political and corporate interests. In the USA, the Committee of Concerned Asia Scholars was formed in 1968 and used its journal *Critical Asian Studies* to publish anti-establishment monographs and bulletins. In 1967, the North American Congress on Latin America (NACLA) was founded, together with its activist bimonthly *NACLA Report on the Americas*. The Middle East Research and Information Project (MERIP) emerged in 1971 and continues to publish its critical *Middle East Report*. Academic activism regarding Africa also emerged at the end of the 1960s. It was not until the late 1970s, however, that a unified voice emerged in the USA. The Association of Concerned Africa Scholars (ACAS) and the African Studies Association (ASA), using its journal *ISSUE*, mobilized against US policy in southern Africa (Wiley 2012: 147–9). In the UK, the first issue of the *Review of African Political Economy* (ROAPE) came out in 1974 (ROAPE 1974). Reflecting the academic autonomy of the time, ROAPE was created to counter the conventional wisdom (today more firmly entrenched than ever) that chronic poverty is an individual behavioural problem that can be reduced by more effective market interaction and penetration. While the journal could provide a forum for discussion and analysis, 'questions of tactics and strategy can be answered finally only by those struggling in Africa itself' (1974: 1).

Rather than dragging anthropology out of its crisis, Marxist anthropologists would deny anthropology a future. Instead, a new practice was needed in which radical anthropology combined with the political struggles of the subordinate and exploited classes to work for 'their emancipation from imperialism and capitalism'

(Banaji 1970: 85). Through solidarity with the oppressed, radical anthropology was a transitional force 'in the struggle of ideas against things' (Buchanan 1973). It was transitional in welcoming its own disappearance as the price willingly paid for a better world.

Fieldwork in Sudan

During the period of decolonization, the international system was relatively open to the circulation of political praxis. As a way of extending the analysis of the relationship between circulation and connectivity, this section looks at the type of structural analysis that was possible during this window of movement. What follows is a reflection on my own ethnographic fieldwork in Sudan during the mid-1970s (Duffield 1981).

Dignity of distance

The original intention had been to complete my Ph.D. fieldwork in northern Nigeria. I had spent the 1972–3 academic year at the School of Oriental and African Studies (SOAS) learning the basics of Hausa. Reflecting the student radicalism of the time, SOAS was a centre of agitation around Third World liberation and anti-capitalist issues. Because of a Nigerian government moratorium on the issue of visas, however, plans changed and I ended up in Sudan. Through a Sudanese friend in London, I had learnt that there were many long-established Hausa-speaking communities in Sudan. They had emerged during the colonial period as a consequence of the overland *hajj* to Mecca. Changing my travel plans and registering as a research student at the University of Khartoum, although all done by letter, proved straightforward. I eventually arrived in the village of Maiurno, in Sudan's old Blue Nile Province, late afternoon on 16 January 1974, a little over three weeks after leaving the UK. I was twenty-four years old and would remain in Maiurno for the next fourteen months.

Anthropology students were expected to live modestly, adapt to local conditions, learn the relevant language and spend at least a year in the field to observe a complete seasonal cycle. Within reason, the object was to blend into the host community. Apart from the advice of supervisors and experienced researchers, unlike today, there were no mandatory methodology courses, upgrade requirements, risk-assessment procedures or ethical committees to satisfy. Access to the outside world had yet to be corrugated by bureaucratic barriers, insurance needs, anxiety-inducing security requirements or ethical uncertainties. The advice from my supervisor was to take gifts and make friends. A box of fragrant soap, cartons of cigarettes, photocopies of historic texts written in the vernacular, and a Polaroid instant camera constituted my attempt. During the 1970s, apart from bureaucratic delays, getting a visa and security clearance for internal travel within Sudan was a routine exercise. Basically, as soon as the grant arrived, placing trust in oneself and others, you set off. Anthropological fieldwork in post-independence Sudan was very much an individual sink-or-swim exercise. In order to succeed, you had to maintain momentum. You learnt, for example, that aspirants who failed to leave Khartoum within a month of arrival would probably never do so.

Maiurno lies about 200 miles south of Khartoum on the banks of the Blue Nile. In the 1970s, a metalled road went only as far as the town of Wad Medani, just over half-way. At the best of times, getting to and from Maiurno was a long and uncomfortable day's journey in an open-sided bus.[5] It was especially difficult during the rainy season when the road south of Wad Medani became a quagmire. It was best to avoid travel at this time of the year. Maiurno was a large, predominantly mud-built village with a central marketplace. Apart from that serving a handful of brick-built compounds belonging to leading merchant families and a renowned spiritual healer (*faki*), there was little electricity. Water was drawn directly from the river, transported by donkeys and sold by the jerry-can. Immunization against the likes of yellow fever, polio or typhoid was normal for research students. However, rather than endless prophylactics or sterilization procedures, the advice was to strengthen one's personal immunity by eating and drinking as the host community did. Local dress was cheap and helped with acclimatization.

Founded at the turn of the twentieth century, Maiurno is one the main centres of *Fellata*[6] settlement in Sudan. My arrival was facilitated by meeting a student from Maiurno then attending Khartoum University. He was one of the first from the village to have done so. With a suitcase and his letter of introduction, I arrived unannounced at the then-dilapidated compound of Sultan Abu Bakr Mohamed al Tahir,[7] where I lived as a guest for the length of my stay, eating with the unmarried men and youths of the compound. Lying awake that first night, listening to the crickets, I felt as if I had travelled to the end of the world. There were no telephones. Even in the UK, outside of the middle classes, companies and public bodies, in the mid-1970s telephone handsets had yet to spread throughout society. Neither my family nor my girlfriend had one. Maiurno was irretrievably at the end of a three-week letter cycle. The only practical way to communicate with the outside world was to open a post office box in the town of Sennar, about 12 miles north. Over dirt roads, just getting there took an hour or so by lorry. The flimsy one-page aerogramme was the main means of international communication. Even with friends in Khartoum, you kept in touch by letter.

Since it took half a day to post and collect mail, the journey into Sennar became a mainly weekly event – not that the journey itself was otherwise wasted. Work expands to fill the time available. Through the circulatory process of posting and retrieving mail, I met people, formed acquaintances and picked up useful knowledge. Going to the post office would become a time for dropping by for a chat, drinking tea, and passing the time of day in Sennar.

Letter-writing was therapeutic and opening the PO box never failed to delight or surprise. However, letters were of little use for sorting out immediate problems or concerns. These would have either gone away or been resolved by the time any supervisor's reply arrived. You dealt with problems directly with the knowledge or contacts at hand. What Peter Sloterdijk (2013 [2005]: 13) has called the 'dignity of distance', a quality properly associated with the days of ocean navigation, still had a legacy existence in rural Sudan in the mid-1970s. This dignity allowed for a level of social immersion that is difficult in today's always-connected world with its distracting ability to be in more than one place at any given time. Today in Maiurno, the telecommunication masts stand taller than the minarets. While the network is slow, you can be emailing within minutes of arrival. There is no more dropping by for tea and a chat on the way to the post office. Instead, there is the habituated pressure to update and expect feedback while continually looking for messages. As an example of the dignity of distance, in the summer months of 1974, I contracted yellow jaundice. Given its debilitating effects, I was put on a lorry and half-carried to Sennar hospital where I was diagnosed and spent a singular five weeks recovering on a public ward. The letter cycle made it possible to finesse this event in a way that avoided unnecessary alarm in the UK. Rather than a setback, such entanglements were an intrinsic part of being there. Such a development today would probably result in a medevac.

Structural method

My original research intention had been to collect contemporary folktales, as told by Hausa speakers, in order to subject them to the structural analysis that Claude Lévi-Strauss had developed in his deconstruction of South American mythological systems (Lévi-Strauss 1968). Influenced by the Marxism of the New Left, the attraction to Lévi-Strauss had been immediate. Uncovering the binary structure of mythological discourse to reveal the underlying human–nonhuman classificatory systems at work delivered, if one was needed, a mortal blow to anthropology conceived as the study of 'primitive' societies. Myths were revealed in their full majesty as the encoded work of meticulous observers and adept analytical *bricoleurs* of the natural environment. While lacking a material infrastructure of comparable density, in terms of nuanced complexity and aesthetic appeal, the mythical logos rivalled the ideological systems of corporate capitalism (Lévi-Strauss 1966). Like a message out of time, Lévi-Strauss' demonstration of this shared ability for critical thought blended naturally with the solidarity of radical anthropology.

During the course of fieldwork, some twenty-five hours of folktales, riddles, histories and songs from a number of mainly rural Hausa-speaking settlements across northern Sudan were recorded.[8] However, while continuing to collect and translate this material, the initial aim to use deconstruction to understand social change was soon abandoned. Once on the ground, questions relating to political economy became more pressing. Disparaged by their Arab neighbours as second-class citizens because of their African ancestry (Duffield 1983), the *Fellata* had been important to the colonial economy as agricultural labourers and peasant farmers. At the same time, the growing impact of commercial agriculture in the Sennar area was evident. Seeing Maiurno in relation to this social and economic background became the primary focus. The methodology, however, was similarly structural. In an interview recorded in 1971 touching on the work of Lévi-Strauss, Michel Foucault defined structuralism as the 'analysis of relations' or 'delineating the relations that there can be among elements in a set' (Foucault 2013: 40). The analogy was given of a photograph of a face. Relational points can be drawn between the main elements of the picture – for example, the distances between ears, nose, eyes and so on. If the positive photograph is now turned into its negative, these elements – in this case, their colour – change. The colours are inverted and it becomes a different picture. However, the relations between the elements remain the same. Despite a colour inversion, a face can still be recognized.

Foucault's earlier works had examined three completely different fields of knowledge that had emerged between the seventeenth and nineteenth centuries; namely, biology, economics and linguistics. The content and knowledge of these disciplines are entirely different and are not comparable. However, 'you can find in each of those fields a certain number of the same internal relations as you can find in other fields' (2013: 41). The way the grammarians isolated words and the mechanism of words, for example, is comparable to how the natural scientists classified different species into hierarchies. To the extent that his analysis demonstrated such underlying relationships, it points to another 'more fundamental and technical point' (2013: 42). Structuralism addresses social realities governed by rules unknown to the subjects who perform them. As a comparative method, it uncovers the unconscious forms of knowledge that Foucault calls the '*episteme* of an era' – that is, the unconscious rules that govern the organization of the various fields of knowledge and experience (2013: 30). Structuralism is thus part of the 'great questioning of the sovereignty of the subject we are witnessing these days in our culture' (2013: 43).

With this in mind, one could ask what distinguishes structuralism from, say, the pattern recognition central to data behaviourism or, as we have already seen, neoliberalism's own questioning of the rational subject with its instrumental assertion of necessary ignorance? In answering, it is important that 'structuralism is the method of analysis that consists of drawing *constant relations* from elements that in themselves, in their own character, in their substance, *can change*' (emphasis added, 2013: 41). Foucault saw this as a characteristic of deductive science generally. It is this assertion of a constant or historic set of relations (the same general *episteme* or structure) existing between elements that constantly change that separates structuralism from the broad sweep of post-humanism. The unconscious relational structures existing in the midst of elemental change are historically given. While individual 'reality' may change, the *episteme* of an era defines and shapes a more stable and enveloping 'world'. Post-humanism collapses this distinction. It disavows any historically given relational structure in favour of the 'complexity' of the pure factuality of environments comprised of constantly changing and emerging elements. In place of history and the world, post-humanism defends the existence of a universal biohuman essence embedded in a never-ending process of emergence and becoming (Dillon & Reid 2000). This is returned to in the next chapter in relation to our changing understanding of famine.

Louis Althusser drew on Foucault's work (Althusser & Balibar 1970 [1968]: 45) in support of his own antihumanist position – that is, 'subjects are constituted through ideological interpolation, and do not preexist such interpolation' (Rouvroy 2012: 12). In other words, in each age, socialization involves imbibing the external and unconscious relational structures that govern the organization of existing knowledge and experience. At the same time, however, the structure that this gives to history and the world creates a space or a commons between the world and individual reality. This space leaves room to contest and critique the governing rules, to understand the power effects of the different knowledges and, in so doing, to take possession of ourselves. In the mid-1970s, the underlying structure that concerned radical anthropology was that of capitalism. While still to gain complete hegemony over society, discovering how capitalism was structuring the Third World was the task at hand.

Being there

Structural anthropology requires the ability to circulate. It requires both the possibility and the commitment to spend time in a place, and to work as close to the ground as possible. This book's Introduction briefly outlined some distinctions between 'knowledge' and 'data'. In relation to the ethos of participant observation that underpins fieldwork, further contrasts can be made. Ethnographic fieldwork allows for interruption and expects delay. It gives people time to speak, to account for and justify their claims. It creates time and opportunities for causal examination, verification and refutation. Data, on the other hand, is only useful if it emerges quickly and is immediately operational. Data is epistemologically remote, turning behaviour into signals and alerts that can be recorded and actioned at a distance. In order to be immediately available, rather than allowing people to account for themselves and their actions, *it avoids engagement* (Rouvroy 2012).

Of central importance in Maiurno was getting to know a social cross-section of the host society. While today the town is less conservative than it was – there being a cultural bar on talking to women, especially married ones, in the mid-1970s – for me there was an inevitable gravitation to the world of men. This meant making the acquaintance of a range of farmers, artisans, lorry drivers, labourers, merchants, *fakis* and teachers; from the highest to the lowest, the educated and the illiterate. Deliberately seeking out differences in opinion, experience and social standing was vital. Apart from entertaining in my room, hours were spent in the fields talking to farmers or labourers, hanging around the lorry stop, drinking tea with market traders or chatting to artisans, as well as passing time in the compounds of acquaintances and friends. Given that the main concern was to find the underlying relations between the main social elements, the methodology involved a patient circulation between them: recording views, experiences and life-histories, at the same time as looking for patterns and relationships within a wider narrative of the development of rural capitalism.

Participant observation cannot be rushed. Ideas and conjectures grow at walking pace, and insights and discoveries worthy of the name reveal themselves slowly. When not visiting other areas of *Fellata* settlement, my normal routine in Maiurno involved a continuous movement between informants, acquaintances, friends and confidants – mostly turning up rather than making appointments. If people were away, or occupied, you moved on. Carrying a notebook and taking notes was essential. Usually in the evening, the day's work would be pulled together, systematized and written-up in duplicate books. The carbon-copies were saved as an original record, while the top-copies allowed a basic level of sorting and classification. The regular write-up of observations, casual conversations and interviews maintained momentum by helping to refine questions, form causal conjectures and uncover gaps that would provide the basis of future expeditions into the life of the village.

These excursions were not exercises in entitlement or expectation. Fluent in Hausa and working alone, it was all about building relationships over time. Respectful of people's wishes and confidences, as much time was spent explaining my world as asking about theirs.

Rough maps, for example, were often drawn in the sand to show unschooled labourers or farmers the position of Sudan in relation to other countries. Marriage and burial customs, together with the nature of work and politics in the UK, were regular discussion topics. My own lack of religious conviction, which was not concealed if questioned, was also a hook for theological debate. The Muslims I mixed with were invariably bemused by my naivety rather than offended or angered. At his suggestion, I made an agreement with one renowned *faki* to explain Marxism in exchange for his views on the psychological dimension of the spiritual healing he practised among the elites of Sudan and Saudi Arabia. After several sessions, how much dialectical materialism I had managed to impart in Hausa was uncertain. I was left in no doubt, however, as to his perceptive skills, grasp of human foibles and general *savoir vivre*.

In the process of gathering evidence, over the fourteen months of walking, talking and exchanging views in Maiurno, some acquaintances became friends, and some friends turned into confidants. Either directly or through their now adult children, these friendships endure to this day. Participant observation morphed into joint adventures within and outside Maiurno. In terms of ushering in a world revolution, the new anthropology failed. Perhaps, it could never have succeeded. Being there, however, led to memorable evenings of camaraderie, jokes and card-playing with the young lorry and tractor drivers who, by temperament, I gravitated towards. Through being there and having time slowed down, they changed my life, just as I changed theirs. At the end of the day, maybe these ties of friendship are what it was all about.

Outcomes

Supplemented by archival research in Khartoum and the UK, the results of this fieldwork questioned a basic tenet of New Left internationalism. By the mid-1970s, an autonomous peasantry had effectively disappeared in northern Sudan (O'Brien 1983). Sudanese independence helped to fertilize the seeds of capitalism that colonialism had sown but, for its own security, had held back (Duffield 1981: 12–13). The power and influence of the Fulani aristocracy in Maiurno had been courted during the colonial period. With independence, this would decline in concert with the integrity of the peasant-based subsistence economy. During the 1950s, large irrigated pump schemes were established around Maiurno. At the same time, the area under commercial rainland cultivation made possible by tractor ploughing began to increase. Given this impetus, wage relations spread through the agricultural economy. Based on large-scale farming and lorry trade along the Blue Nile, a new commercial elite emerged.

In the mid-1970s, two opposing but interconnected tendencies were observed. Namely, 'for the economic function of the family to break down among the poor whilst it [was] transformed and maintained among the rich' (1981: 159). Being dependent upon household labour, the commercialization of agriculture had worsened the terms of trade among small farmers and weakened family ties. Young men sought escape through labour migration or were drawn into seasonal employment on the surrounding mechanized farms. The rich and better-off benefitted in many ways from this process of dissolution and proletarianization among the poor and disadvantaged. Among the former, family relations consolidated around a range of interconnected agricultural, professional and mercantile activities that were divided between sons, brothers and relatives.

As well as marking the peak of Sudan's liberal hour, the mid-1970s was, in retrospect, a high point of post-colonial internal rural development. Although marked by emerging divisions and inequality, like the welfare-Fordism that was then reaching its zenith in Europe and the USA, the relative security of the period is often invoked with some nostalgia among older Sudanese (Bannaga 2002). By the end of the decade, things were changing. As a result of growing debt dependence and increased donor leverage, Sudan would open its agricultural economy to the competitive forces of the world market. The transition of Fordism to post-Fordism and the reconfiguration of the corporation as networked global factory was accompanied by structural adjustment in Sudan, a drop in living standards and a growing informalization of the economy. Formally autonomous peasants were fast becoming a dependent precariat.

In 1977, the construction of a road was begun, starting at Wad Medani, linking Khartoum to Demazin in southern Blue Nile. Within a few years, this all-weather road would run within 100 yards of Maiurno. On unpaved roads, Demazin had been a gruelling day's journey to the south of Maiurno. It now became a matter of hours. This opening up enabled mechanized commercial farming to extend further south, helping to accelerate the dissolution of peasant agriculture at the same time as introducing outside competition into what had been a locally controlled lorry trade. When the rekindled civil war in South Sudan engulfed the southern Blue Nile region in the mid-1980s, this already-declining business, once an important means of wealth creation in Maiurno, suffered a fatal blow (Abu Manga 2009). Within a decade, the Maiurno uncovered through fieldwork, a place where the dignity of distance still had a legacy presence, was a fading memory. Sudan was being connected into new flows of capital, polarized by Islamist resistance and, importantly, it was becoming known through new forms of humanitarian knowledge introduced by the NGO-led aid invasion of the mid-1980s.

Postscript

In 2014, almost forty years to the day since first arriving, I returned to what is now better described as the small town of Maiurno for a short retrospective study (Duffield 2014). In the mid-1970s, I had entered through the front door, so to speak. However, due to the ground friction of barriers, bureaucratic hurdles and institutional requirements that now corrugate and striate international space, returning would have been difficult, maybe impossible, without being able to go around the back. Having recently retired, there was no need to navigate the now-mandatory university risk-assessment process, satisfy ethical guidelines, follow private security advice or comply with insurance requirements (Jaspars 2015: 23–4). As in the 1970s, I travelled without insurance – or, at least, decided to forgo the time and complications of trying to get it. Being able to self-fund travel costs meant avoiding the competition for research grants at a time of increased management prescription. The complications of awaiting sabbatical leave, again more restricted than before, were also avoided.

Given the history of Western sanctions against Sudan, obtaining an entry visa for academic research is no longer straightforward. Fortunately, an old friend provided a letter of invitation from the University of Khartoum. For international aid workers or researchers, getting travel clearance to leave Khartoum can be a lengthy and often fruitless process (Jaspars 2015). A longstanding hotelier friend obtained a general tourist travel permit that, centred on Khartoum, covers the main areas of tourist interest. While its validity was then being challenged in the areas of archaeological interest to the north, fortuitously, the permit also extended to the less-travelled south as far as Sennar near Maiurno. Unlike in the 1970s, there is now a visible police and security presence in the village. Early on my first morning, I was detained by plain-clothes security and taken back to Sennar for questioning. Fortunately, I had lived in the same compound as the present *Omda*, or local representative, when he was a teenager. He was able to intercede and vouch for me. With this assurance, I stayed for the next five weeks in the hands of my surviving friends. Given the ground friction that now exists, going through the expected channels would probably result in never getting there, let alone being granted leave to stay. However, if you did make it, you can be checking emails and updating social media pages within minutes.

The following chapter focuses on the NGO invasion of the 1980s. In terms of growing remoteness, of particular importance is how the new aid-based ontologies and methodologies would quickly negate the area expertise, language acquisition, structural methods and forms of ground engagement described above. This invasion would help to prepare the ground, so to speak, for the computational turn that accelerated from the end of the 1990s.

Chapter 5

FANTASTIC INVASION

In the mid-1980s, borrowing from Dan Large (Large 2012), northern Sudan was the site of a *fantastic invasion*.[1] Having little or no presence before, there was a rapid flush of international NGOs in response to the drought and famine conditions of the time. Apart from a small UN presence, and those NGOs that appeared in South Sudan following the end of the first civil war in 1972 (Tvedt 1994), there were few aid agencies operating in the North before 1983. Within a couple of years, over 100 NGOs, of varying specialism, capacities and ground presence, quickly materialized (Jaspars 2015). The British NGO Oxfam, for example, began working in Sudan in a small grant-giving capacity oriented to promoting a community self-help model of development. During 1984, however, benefitting from the media interest generated by the drought also affecting neighbouring Ethiopia, this quickly grew to become Oxfam's largest humanitarian relief operation at the time (Walker 1987). Modest by today's NGO standards, it had an annual budget of over £1 million and employed around 200 people, more than 90 per cent of whom were Sudanese.

At the end of 1985, together with my family, I returned to Sudan as Oxfam's Field Director. Based in Khartoum, the programme covered both northern and southern Sudan. We left a few months before the Islamist coup in June 1989. The job description was to ease the Sudan programme from its humanitarian footing back to a community development focus. Given the effects of the rekindling and escalation of civil war in the South, however, if anything, the operational character of the programme increased.

The 1980s were a period of rapid expansion for international NGOs generally.[2] In relation to Sudan, the term 'fantastic invasion' is entirely apt. Creating employment for many Sudanese along the way, it was rapid and irrevocably changed things. Taking place within the space of a few years, old ontologies, expertise and methodologies gave way to new beliefs and aid practices. The invasion was disorientating in terms of its rapid displacement of the structural analysis and ethnographic methodology that had been in vogue a decade earlier. Area-specific knowledge and local expertise derived from language began to disappear in favour of faster, more immediately actionable and transferable grids of information. The energy of the anti-capitalist May '68 critique had been recouped as a progressive neoliberalism based upon an assertive cosmopolitanism that combined rights and gender equality with the twin aims of direct humanitarian action and making markets work for the poor.[3]

This and the following chapter analyse this important period of anticipation, trialling and rupture. Having a different emphasis, they both examine the ontological and methodological prescience of the fantastic invasion with regard to the consolidation of the cybernetic *episteme* and the coming post-social world. This chapter begins with an overview of direct humanitarian action. During the course of the 1980s, the political and organizational autonomy necessary for this would narrow and close. However, it provides a contrast with the risk aversion of today. Importantly, direct action also relied on human agency and having a ground presence. Chapter 11 analyses how the indignation that drove direct action has now been transferred to the self-acting humanitarian object. The rest of the current chapter examines the anticipatory nature of the new ontology of disaster the fantastic invasion helped to develop. The shift from modernism to postmodernism is explored in relation to the changing fortunes of the refugee camp and a cybernetic reading of Foucault's understanding of liberal security. The final part of the chapter considers the emergence of complexity-thinking in relation to famine, and the replacement of the explanatory power of history and causation by that of environmental uncertainty.

Direct Humanitarian Action

As Boltanski and Chiapello (2005: 350) have perceptively argued, part of the recoupment of the May '68 critique involved its displacement into direct humanitarian action. By the time of the fantastic invasion, the network metaphor as a means of describing the social world had already displaced structural concepts like social class. Class and exploitation had been rendered obsolete by the praxis of inclusion/ exclusion (2005: 349). The 'excluded' were those now exterior to the system, 'without a voice, without a home, without papers, without work, without rights, and so on' (2005: 349). Reflecting this emerging register, in 1971, at a time when the opportunity for world revolution was already seen as waning (Marcuse 1972: 2), a group of young French doctors established Médecins Sans Frontières (MSF) 'to bring down all the barriers, all the borders that still stand between those who have the vocation of saving' and the victims of conflicts and natural disasters, especially in the developing world (MSF 2017). These activists had turned their backs on what was now seen as the ineffectual class-based political struggles then taking place in France. Rather than revolution, they promoted instead an activism directed towards the humanitarian victim that, as the epitome of the vulnerable and excluded subject, demanded mediation, protection and an end to suffering. It was an activism that placed its emphasis 'on engaging in action and direct individual aid' (Boltanski & Chiapello 2005: 350).[4]

At a time when the international plane was still relatively open to the circulation of political praxis, NGOs enjoyed a degree of autonomy that is absent today. During the Cold War, development assistance was a tool of Cold War foreign policy, especially in servicing anti-Soviet alliances in the Third World. Importantly, however, it was also a time when it was taken for granted that 'NGOs can operate in countries their governments do not favour' (Whitaker 1983: 49). Reflecting this independence, NGOs often found themselves opposed to the foreign policy objectives of their governments. In Cambodia, Vietnam, Ethiopia and Nicaragua, for example, international NGOs waged public campaigns, including the independent delivery of humanitarian assistance, against government policy (Duffield 2007: 54). This autonomy persisted into the 1980s. The mediatized launch of Band Aid in November 1984 was its last gasp for air. Focusing on the famine in Ethiopia, Band Aid provided the media backdrop in Europe and the USA to the fantastic invasion. Initially funded through the proceeds of the chart-topping single 'Do They Know It's Christmas?', Band Aid's finances were further bolstered the following year by the Live Aid concert broadcast on

TV from London and Philadelphia. Catching politicians on the back foot, Band Aid struck an indignant, direct-action pose against Cold War political conventions, diplomatic niceties and bureaucratic delays that were stopping help reaching the famine victims in the Horn of Africa. This indignation is well reflected in an outburst by Bob Geldof: 'I'm not interested in the bloody system! Why has he no food? Why is he starving to death?' (Geldof 1985).

The fantastic invasion saw the rapid appearance of NGOs on the ground in places like Sudan when field staff enjoyed a residual degree of autonomy. Not only was the Cold War balance of forces favourable – even if funding came from government, the subcontracting relationship between donors and NGOs was then relatively arm's-length. The underdeveloped state of telecommunications, moreover, continued to provide a relative 'dignity of distance'. Such factors gave international aid workers on the ground more independence than they enjoy today. The Oxfam action-oriented title of 'Field Director', for example – which would soon be changed to a more corporate 'Country Representative' – echoed the old system that gave the person on the spot significant programme decision-making authority.[5] Historically, there had been an element of necessity in this dispensation, due to delays in communication. In the mid-1980s, the available telecommunications in Sudan were still largely based on pre-World War II technology. NGOs in Khartoum relied on Telex[6] to communicate with their HQs. The only reliable link for the dozens of agencies then operational was through a city-centre hotel that maintained a back-door access to the main Post Office telephone exchange.[7] Supplementing a regular couriered mail pouch, internal communication between Khartoum and Oxfam's field offices was by short-wave radio. All communication between the field offices and the HQ in Oxford went through Oxfam's Khartoum office. Getting outline updates from field offices could take several days.

For several years, this limited connectivity kept a residual autonomy at field level. It also coincided with a period when Western donors and UN agencies were politically aligned to the Sudanese state. Still in the thrall of Cold War thinking, they were complicit with the efforts of the military and security services to blockade the South and starve the rebels into submission (Keen 1994). For NGOs willing to take it, there was an opportunity to target the complicity between the international community and the Sudanese state. Compared to the comprehensive security protocols and risk-avoidance that characterizes the present aid encounter (Collinson & Duffield 2013), during the 1980s international NGOs travelled relatively freely within Sudan's famine- and war-zones. NGOs managed, for example, a low-key cross-border relief operation into the rebel areas of northern Ethiopia and Eritrea (Duffield & Prendergast 1994). Oxfam ran its own ground-based monitoring missions into these areas, with international staff travelling in a Land Cruiser that had been hand-camouflaged as a gesture of protection against Ethiopian Migs. In several besieged government towns in South Sudan, aid agencies supported humanitarian operations for war-displaced groups. Getting into these towns, surrounded and periodically under threat of being overrun, involved privately chartered light aircraft corkscrewing down onto the runway to minimize the threat from ground-to-air missiles. How one left depended on whether the pilot was a fast-and-low or corkscrew-up person. These flights continued even when aircraft were occasionally brought down. Inconceivable today, such risks were an accepted part of direct humanitarian action.

Having people on the ground in disaster-zones, and access to independent funding such as that provided by Band Aid, gave NGOs a political advantage over donor governments and the UN. This mobility differential translated into the capacity to collect, analyse and disseminate verifiable information. With one or two exceptions, donor representatives and diplomats had no direct field presence. Having links to the media and human rights groups, NGOs were able to play donors off against each other, threaten to embarrass governments, use the media to their advantage, pass evidence to human rights groups, and exert pressure on uncooperative UN agency heads.[8] Direct humanitarian action involved leveraging the NGO mobility differential to expose the complicity between the international community and the Sudanese government while widening humanitarian access. Humanitarianism often lays claim to an idealistic history reaching back to the civilizing struggles of the nineteenth-century Europe (Barnett 2011). During the 1980s, however, the above were its concrete conditions of existence. Moreover, this window of opportunity would close with the ending of the Cold War.[9]

This brief account of direct humanitarian action should not read as nostalgia or romanticism. The autonomy on which it was based was abused more than used by aid agencies and practitioners. Moreover, as argued in the next chapter, the mobility differential is, essentially, a mechanism of exploitation used by mobile NGOs to extract some political, cultural or career advantage from immobile beneficiaries. While it is possible to argue that humanitarian action was a more principled use of the mobility differential than the paternalism of developmentalism, they were both bound up within the progressive neoliberalism of the fantastic invasion.[10] In this respect, as argued in chapter 11, the indignation that drove the direct humanitarian action of the 1980s has been incorporated within the automatic and remote technologies of today's humanitarian innovation. The rest of this chapter is devoted to the more direct anticipatory legacies of the fantastic invasion.

New Ontology of Disaster

Until the mid-1980s, as an object of social knowledge, Sudan had been defined by historians, political scientists and anthropologists. In fairly short order, the fantastic invasion undermined the modernist and structural assumptions that had sustained this academic hold. In particular, it introduced a series of ontological, epistemological and methodological disruptions that replaced historical understandings and thick descriptions with an information-dependent cybernetic mode of thinking based on complexity and behaviourism. Since this move from knowledge to the production of data took place in advance of the computational turn and the machines necessary to realize it, this preparatory stage is called 'ground work'. While complete in itself, and associated with the aid regime of livelihood support which is examined below, this ground work drew upon a new ontology of disaster that challenged earlier modernist assumptions.

Modernism saw disasters as accidental or unforeseen events that were external to the normal workings of society. They were abnormal or extraordinary occurrences. Although they could be devastating, society would eventually return to normality. Because they were external to society, however, this meant that steps could be taken to protect against them. These included better scientific prediction as well as measures to contain and shield society from their effects – for example, investing in volcanology or similar sciences at the same time as initiating large-scale engineering works, like dams or canals; evacuating or transferring at-risk populations; establishing quarantine regimes; mass vaccination programmes; or opening and managing temporary camps for the disaster-affected. The modernist ethos was to protect against disasters, including predicting and containing them, while walling off society from their effects (Hewitt 1983).

Although it would be several decades before the term 'resilience' became a *lingua franca* across the natural and social sciences, in 1973 Crawford Holling published his celebrated article challenging the modernist approach to managing ecological systems. In particular, protective management was argued to have negative consequences because natural species survive and adapt better poised on the edge of extinction. During the same decade, the modernist approach to natural disasters also began to be questioned. Rather than something accidental or outside normal society, disasters were increasingly seen as internal or integral to the working of society. Disaster lurked within the process of social change, the structure of the built environment and the contrasting registers of vulnerability and exposure that defined different social groups. By the end of the decade, the very notion of a 'natural disaster' was being questioned (Wisner et al. 2004).

Empty the camps

For want of an alternative, refugee camps remain, and will continue to remain, an enduring necessity. In recent decades, however, they have attracted cumulative negative connotations. As the international asylum regime has shifted from protection to exclusion (Hammerstad 2014), refugee camps have been evacuated of any developmental content. Having been progressively securitized, in many challenging environments they have evolved into international no-go areas (Agier 2011). During the 1970s, however, the refugee camp was still seen as a positive humanitarian technology. In Africa, the camp exemplified the modernist approach to disaster. As a self-contained managerial solution, camps separated refugees from the surrounding society. They attempted, quite literally, to wall off the politics that had been responsible for the refugees' forced displacement. The 1969 Refugee Convention of the Organisation of African Union (OAU) is important here. In a new move in refugee law adapted to African political conditions, the Convention conferred a formal humanitarian status on refugees, including asylum recognition, documentation and support from UNHCR, in exchange for their temporary suspension of politics (Karadawi 1999). As long as refugees kept to the internationally recognized space of the camp and its environs, and renounced political activity while under the auspices of UNHCR, they were eligible for humanitarian protection and assistance guaranteed by the host government. The ability to demarcate spatially and transform an active political subject into a compliant humanitarian one was the cornerstone of traditional humanitarian neutrality. Being able to quarantine politics, as it were, was central to a humanitarian conception of the prisoner-of-war as well as the refugee.

In this respect, Giorgio Agamben is right to argue that the fire-wall which neutrality places between itself and politics produces a humanitarianism that maintains 'a secret solidarity with the very powers [it] ought to fight' (Agamben 1998: 133). However, it is also the case that modernism did not deny the political or seek to suppress its existence. Rather, it attempted to contain it within a defined and collectively recognized field of action. In distinguishing modernity from data behaviourism, Antoinette Rouvroy (Rouvroy 2012) sets out a number of characteristics of the former which can be seen in the refugee camp. The camp functioned spatially rather like a court of law. Or, at least, it created a recognized break or formal interruption in the flow of events. It opened an institutional space where politics could be suspended, norms renegotiated and subjects allowed to account for themselves through language. Provided the politics of displacement were suspended, no judgement was made regarding true beliefs, or future intentions; an inner private space was allowed. Unlike computer profiling, whereby no one escapes their recorded history, the structure of the camp allowed a subject to circulate between the political and the humanitarian, the past and the future. As discussed in chapter 10, with the securitization of the camp and increasing reliance on remote satellite sensing and data informatics, refugees are now understood ecologically as integral parts of their natural and logistical environments. In what is seen by some as a celebratory development (Slim 2015: 595), they are also being biometrically tattooed (Agamben 2008) so that their transnational movements and entitlements can be remotely tracked.

Inflected with the anti-hierarchy sentiments of the May '68 critique, as a modernist institution, the refugee camp, like food aid, entered a sustained period of progressive criticism from the end of the 1970s (Harrell-Bond 1998). Critics argued that top-down management regimes created dependency and stifled innovation. Rather than affording a distinction between political and humanitarian identities, separation from society operated to prevent refugees from integrating and using their skills and experience for self-improvement. Camps also created parallel systems of service provision that produced destabilizing imbalances between refugees and local groups, and eroded government capacity. In short, refugee camps worked against the progressive neoliberal aim of making capitalism work for the excluded. Camps stood in the way of market integration, preventing refugees from exercising their freedom of choice and taking responsibility for their own self-management.[11] During the course of the 1980s, the refugee camp was ideologically emptied of any positive content by an assertive cosmopolitanism.

As a modernist construct, the refugee camp existed at a time when states could offer refugees an alternative. They proffered the possibility of transforming an exceptional humanitarian identity back into an accepted one within the parameters of national development planning. Symptomatic of the transformation of Fordism to post-Fordism, progressive neoliberalism's ideological emptying of the camp reflected the privileging of 'community' over and above state provision. In the global North, this positioning was part of the death of the social (Rose 1996). In the South, where state provision had been limited and usually concentrated in urban areas, elevating the community was anticipatory regarding the exploration of new post-social relations and subjectivities. Community 'development', 'empowerment', 'participation' and 'ownership' were all key slogans of the fantastic invasion. As we shall see, however, the current resilience regime foregrounds the 'personal' ahead of the collective. Despite this displacement, during the 1980s and 1990s, as a construct of mutuality, authenticity and self-reproduction, the 'community' was an ideological heavy-lifter in the preparatory ground work for the computational turn.

The privileging of community also called forth a new apparatus of security. Empty the camps, metaphorically speaking, and you remove the wall that modernism placed between society and the politics of forced displacement. The camp sought to contain the political within a set of state-sanctioned parameters. If you remove these barriers and integrate refugees within the community, the political necessarily expands to envelop the whole of society (Agamben 2005). These postmodern shifts, however, are often associated with the opposite – with claims that we have entered a post-political age (Fukuyama 2002 [1989]). On the contrary, if the political expands, security is likewise amplified to embrace the whole of society. This situation is better understood as the advent of the hyper-political (Andreotti 2014). As Marcuse feared (Marcuse 1972), we now have an unregulated governmental urge to leave nothing hidden or undisclosed. In the process, any sense of entitlement to private space, independent autonomy or place of refuge is eliminated. Indeed, to claim or defend such autonomy is now itself a cause for concern. The will to secure against any form of vulnerability or weakness, from famine to terrorist radicalization, would henceforth be predicated on the surveillance, capture and algorithmic analysis of the signals and alerts thrown off by behavioural change across the entire spectrum of social and economic activity, from the price of goats (UNGP 2009) to a person's in-cabin meal preferences (Amoore 2011). Foucault's essentially cybernetic understanding of liberalism as an apparatus of security captures the hyper-political realm.

The steersman

Foucault gave his celebrated outline of liberalism as a biopolitical apparatus of security in several lectures at the Collège de France in 1978 (Foucault 2007). Here, these lectures are read as reflecting contemporary developments – namely, the new ontology's bringing of disaster from outside to inside society and, at the same time, the anticipation of the cybernetic nature of the post-social security apparatus then coming into existence (Deleuze 1992). The lectures also indicate how circulation is the embodiment of insecurity, while security is a technology of connectivity.

The idea that human beings are a biological species became an object of political strategy during the eighteenth century. The resulting biopolitics is not concerned with the legal subjects of sovereignty or with people, groups or communities that can be disciplined into certain behaviours. As a security mechanism, biopolitics emerges in relation to the sphere of circulation between people and things. It is associated with the need to open up previously closed towns and restricted guilds to unimpeded circulation and exchange. In other words, it emerges alongside capitalism. As well as benefits, however, freeing circulation can also have negative consequences. Opening commercial exchange between towns also eases the movement of vagabonds, thieves, seditious ideas and disease. Through surveillance and policing, security seeks to prevent bad circulation as a necessary condition for maintaining the good. Even with surveillance, however, unimpeded movement is still risky. Like entropy, dangers can never be wholly nullified within an open circulatory system. Some problems and misfortune are inevitable. Under these circumstances, unlike discipline, which either allows or prohibits, liberalism weighs the risks and consequences through a risk-based security apparatus that 'works on probabilities' (Foucault 2007: 19).

For liberalism, the future is not exactly controllable or precisely measurable. Governments have to contend with what might happen. Security emerges in relation to the uncertainty and unpredictability of circulatory events and effects.[12] The terrain of security, which is also the space of circulation, is the milieu.

> The milieu is a set of natural givens – rivers, marshes, hills – and a set of artificial givens – an agglomeration of individuals, houses, etc. The milieu is a certain number of combined, overall effects bearing on all who live in it. It is an element in which a circular link is produced between effects and causes, since an effect from one point of view will be a cause from another. (Foucault 2007: 12)

Overcrowding, for example, encourages disease, which creates more dead bodies, and more bodies further exacerbate disease. As Foucault himself points out, the term 'milieu' was not current in the eighteenth century. Describing a synthesis of human and nonhuman systems based on the exchange of information between them, 'milieu' speaks more to the contemporary cybernetic *episteme* (Lafontaine 2007). The human and nonhuman 'natural givens' within a milieu are connected by a 'circular link' between causes and effects. In other words, they are connected by information feedback loops.

Security as a calculus of risk is illustrated in relation to the problem of famine in eighteenth-century Europe. The disciplinary mechanisms then available to government tried to prevent famine by keeping grain prices low through measures like price and export controls, prohibiting hoarding or enforced changes to cultivation patterns. In the main, however, such interventions were counterproductive, often acting to increase prices and exacerbate shortages. In shifting from trying to prevent famine to managing it, a new liberal apparatus of security emerged. Rather than control, the emphasis was on liberalizing domestic and foreign exchange. It was observed that allowing the free circulation of grain not only maintained profit – the same unimpeded movement also was 'a much better mechanism of security against the scourge of scarcity' (Foucault 2007: 34). *Laisser-faire* and *laisser-passer*, or letting things move freely and take their course, came to define the liberal way of managing the insecurity caused by scarcity (2007: 41). In contrast, discipline is a centripetal force: it tends to concentrate, focus and enclose. It allows nothing to escape or run its course; discipline either permits or forbids. Security, on the other hand, is centrifugal and expands by integrating ever more new elements within itself. Security stands back, allowing circulation to take place, while remaining alert and trying to make the existing arrangements work (2007: 48–9).

In the lectures, Foucault contrasts security with discipline. For us, the suggested relationship between circulation and connectivity is also important. In the famine example, security connects distant markets, brings areas of surplus and deficit into exchange, and aligns producers and customers. Security helps to create the capitalist world by refashioning the economy as an interconnected global market. Security is a technology of connectivity within this world. While the market does not eliminate scarcity or famine, it helps to manage it. Liberal freedom, or the ability of people and things to move, the condition of their changing place, and of their circulation and exchange, is the necessary correlative of the deployment of interconnecting apparatuses of security. This is a cybernetic conception of security (Lafontaine 2007).

As Giorgio Agamben (Agamben 2013) has pointed out, there is a resonance between the Greek etymological origins of the term 'cybernetics' and Foucault's analysis. 'Cybernetics' has roots in the term *kybernan*, meaning to steer or govern, and its corollary *kybernētēs* – that is, steersman or governor (Wiener 1954: 15). The control of a machine in accordance with its actual – as opposed to expected – performance is made possible by the operation of feedback. As a cybernetic agent, the steersman of a boat masters the circulatory nature of the sea by transforming the sensual feedback from the wind into the adjustment of the sails and tiller for the purpose of defying shipwreck through the connectivity of sailing. The steersman does not resist the sea in a disciplinary or modernist sense – that is, by trying to prohibit the waves, or by building massive harbours or constructing Promethean dykes; he goes with the circulatory currents, riding the waves and bending the wind to achieve desired ends. Reflecting Foucault's analysis, the steersman stands back and lets things happen. His skill translates into the 'ability to govern and guide them in the good direction once they take place' (Agamben 2013).

The notion of biopolitics sits awkwardly in this schema. During the course of the nineteenth and twentieth centuries, a growing organized resistance to capitalism emerged among the people. Encouraged by this resistance, a biopolitical apparatus of security was inserted within the circulatory milieu of interconnected human and nonhuman systems. This took the form of normative urban planning, public health, mass education, social insurance and organized policing (Rabinow 1995). While these security measures maintained circulation, by the mid twentieth century, in the form of a social democratic welfare state, this apparatus was not, so to speak, sitting back and letting things happen. Its bureaucrats, regulations and social insurance schemes intervened between populations and market forces in an attempt to protect the former from the latter. What we appear to have in the steersman is a pre-social apparatus of security that, in the post-social conditions of the new economy, has now come of age. We can join Étienne Balibar in his suspicion that, perhaps, 'we are only now entering capitalist society ... or, if you prefer, we're only now entering "pure" capitalism, which does not have to deal constantly with heterogeneous social forces that it must either incorporate or repress, or with which it must strike some form of compromise' (Balibar 2016: 22).

As Marcuse feared, decades ago, the new economy has only to deal with the effects of its own logic of accumulation and the things necessary for its reproduction. Dispensing with norms and averages, its subordination of the individual to the data profile traces the assertion of capitalism's hegemony over society as a whole, including its previously autonomous psychological, social and reproductive spheres. The singularity of this hegemony is that, in its completion, capitalism has been freed from any social responsibility. The steersman does not represent the death of biopolitics. It signals a move from a herd-like species-being associated with a statistical or normative understanding of population to embrace the leaner, more agile and personalized aesthetics of a biohuman predator. Moreover, the steersman is no longer human but comprised of countless algorithms working through the networks of computational technologies, platforms and data-parks that make algorithmic governmentality a possibility (Rouvroy 2009). Capitalism now connects everything on a planetary scale, from smart cities to the most remote deserts; every house, refrigerator, mobile phone, wrist watch, electricity meter, TV or car. The internet interconnects and exchanges information between people and things as never before. Through smart phones, social media and recording devices in homes, people share their moods, opinions, leisure activities, food habits and sexual lives in real time. As the Organon of security, with the arrival of Big Data the steersman now stands forth (Rahebi 2015). In helping to trace the genealogy of this emergence, the next section examines the appearance of complexity-thinking in Sudan during the 1980s.

Theory to Complexity

When I returned to Sudan at the end of 1985, many of the socialist and left-leaning Sudanese academics that I had known at Khartoum University six or seven years earlier had either left for Europe or the USA, or else lowered their profiles. The introduction of Sharia Law the year before, and the rekindling of war in the South, marked the end of Sudan's liberal hour. Although the academic study of capitalism was all but over, a structural approach to the existing famine conditions still had a legacy presence. Works like Jay O'Brien's (1985) 'Sowing the Seeds of Famine' or Taisier Mohamed Ahmed's (1989) *The Cultivation of Hunger* argued that the famine had been exacerbated by the unregulated expansion of commercial agriculture. What amounted to the mechanization of agricultural strip-mining had intensified the crisis among subsistence farmers and pastoralists through land dispossession, at the same time as disrupting nomadic transhumance. Dispossession and restriction, moreover, created an expanding pool of casualized, cheap and disposable labour that was being deposited in urban slums or else put to work on the new commercial farms. When combined with poor rains, the degraded subsistence economy had been unable to resist the spread of famine conditions. Reflecting the central argument in Fatima Mahmoud's (1984) *The Sudanese Bourgeoisie*, the wealth accruing from this process of dispossession and exploitation was concentrated within the integrated riverain kinship networks of Sudan's political and commercial Arab elite.[13]

Now in decline, this New Left internationalism expressed an essentially political, rather than a technical or operational, encounter with disaster. Drawing on the legacy of Third World activism, the logic of the analysis invited political reform, including the break-up of elite control, the regulation of commercial agriculture, rural representation, labour protection and curbs on dispossession. Inflected with the new radicalism of progressive neoliberalism, the fantastic invasion had difficulty digesting this agenda; it did not translate into the practitioner demand for information that was directly operational within a stance that now accepted and had made peace with capitalism. Published in 1985, David Booth's 'Marxism and Development Sociology: Interpreting the Impasse' sought to give the fantastic invasion an ideological legitimacy in relation to the albeit waning challenge of structuralism and metatheory (see also Edwards 1989). The core of this conservative countermove was to assert that the empirical 'complexity' of people, things and environments undermined the possibility of structure. Given the huge and observable diversity that exists, how could there be an underlying structure or general causation? It just didn't make sense: 'the complex and challenging issues of development in the Third World today cannot be sufficiently grasped in terms of the dynamics and differential spread of the capitalist mode of production, the theoretical primacy of which is the hallmark of the "classical-Marxist" approach' (Booth 1985: 776).

With accusations of tautology, circularity, conflicting empirical evidence and functionalism, the thrust was to privilege a narrow empiricism of multiple and heterogeneous internal factors. Rather than history, the potential of existing relations to change was more important. Capitalism disappeared, lost in multiple local and conflicting variations; things were system-specific and non-generalizable: 'The practice of development work teaches us that problems are often specific in their complexity to particular times and places. A number of recent studies have recognized this in their exploration of the complex relationships which evolve over time between people and their environment within geographically-restricted areas' (Edwards 1989: 120).

The economy assumes no single pattern or form across societies or countries. There is no 'necessity' to anything and the observable diversity of things undermines any claim to 'general laws' (Booth 1985: 773–4). Moreover, and this was the rub, such theoretical abstractions were divorced from 'real world problems' (1985: 777). And it was real-world problems that the fantastic invasion had come to resolve.

By the beginning of the 1980s, reflecting the conservative counter-revolution then gathering momentum, rather than doing 'theory', the social sciences were embracing the practical challenge of making themselves relevant to the new world of 'policy'. Having lost its protective enclosures and modernist quarantine measures, in the emerging post-social world, the whole of society was now a terrain of security and open to new non-governmental forms of engagement.[14] Focused on real-world problems, the 1980s saw the rise of the practitioner who, as a historical figure, would become increasingly important as politics changed into policy. The practitioner was a central character in the implementation of the NGO-led livelihood regime then taking shape in Sudan. It is here, moreover, that we can see how complexity-thinking changed our understanding of famine.

Environmental exposure

The concept of a 'complex emergency' first entered formal UN usage in Mozambique towards the end of the 1980s, and was soon being applied to describe the situation in South Sudan (Duffield 1994: 9). It was further popularized within the aid industry following the formation of the UN's Department for Humanitarian Affairs in 1992. Complexity-thinking, however, was already integral to Sudan's fantastic invasion during the mid-1980s. As a way of examining the cybernetic *episteme* underpinning the fantastic invasion, Nick Cater's (1986) influential report for Oxfam, *Sudan the Roots of Famine*, offers a starting point. In terms of the medium being the message, the report was then indicative of a new trend that, by the beginning of the twenty-first century, had already become the dominant form of public written communication – that is, the glossy infographic where text is minimized or bullet pointed and subordinate to graphs, diagrams or photographic material.[15] In the case of *Roots of Famine*, text is broken up and punctuated with sidebars, dialogue boxes and photographs. As if designed for neoliberalism's *Homo inscius*, its layout combines simplified and signposted key arguments with visual props to maximize cognitive and emotional impact. Instead of a causal analysis, poverty and hunger are presented as an indeterminate 'complex combination' of many different social and environmental factors (1986: 1).

The term 'complexity' not only deters speculation on a possible structural relation between these factors, but, reflecting its progressive neoliberal pedigree, also signals a world beyond proper human understanding where innumerable factors, conditions and transactions collide. While all bringing something different or special, none can be assigned any overriding importance. In the *Roots of Famine* report, drought, civil war, colonial legacies, cash-cropping, new agricultural methods, unresponsive political systems, closed economic frameworks and gender inequality, in no particular order, all vie together in driving food insecurity.[16] While no main cause predominates, however, there is a core argument. Individually, or severally, such factors can usually be tolerated, but when acting in concert: 'they damage the old ways of life and erode the networks of support that existed within families, villages and tribes. As sustainable food production is abandoned, the intricate mechanisms of survival break down and the common climatic variation of drought ceases to be an endurable calamity and becomes an engulfing disaster' (1986: 1).

While 'complexity' may have a functional role in physics or mathematics, when applied to society as a way of signalling indeterminacy, despite attempts to suggest otherwise (Urry 2003), it explains little or nothing. On the contrary, it deters curiosity and rationalizes ignorance. Other than outlining a range of different contributory factors, no determinate explanation can be given to what involved thousands of deaths in Sudan. While the Oxfam report agrees with the structural analysis that Sudan's subsistence economy was in a state of crisis and dissolution, the complexity argument generates no critique. By not admitting to a world beyond the multiple realities of the different contributing factors, it effectively decapitates theoretical praxis. This is more than negligence or oversight; it has a positive paradigmatic function.

When one examines what the affected 'families, villages and tribes' are opened or exposed to as a result of this erosion of networks, the difference between a structural and complexity perspective is significant. Stripped of the protection of a subsistence economy, for the former they are expelled and made available for historically given forms of exploitation consistent with continued capitalist expansion. As will be argued in chapter 8, what was happening in the 1980s was an acceleration of the informalization of the economy, the growth of urban slums and the emergence of a global precariat. For NGOs, however, something different, even magical, was happening. Admitting to no structure or causation, history and society melted away to be replaced by a *hostile environment*. Rather than being incorporated within an expanding set of capitalist relations, as the 'intricate mechanisms of survival break down', the poor were instead being exposed to the post-social world of the Anthropocene. As a term denoting a distinct geological epoch in which human actions are seen as having irreversible environmental effects, the Anthropocene would not be popularized until the 2000s (Chandler 2018: 2). However, in the sense that the environment was now striking back, as it were, it was already anticipated in the fantastic invasion. Michael Edwards, then Oxfam's Regional Director for Southern Africa, in drawing evidence from Zambia on the relationship between nutrition and cash-cropping, argued that there exists, 'a tremendous diversity in the informal networks of exchange and innovation that have evolved over hundreds of years to ensure peoples' survival in the face of hostile environmental conditions' (Edwards 1989: 120).

The main danger resulting from the dissolution of subsistence farming is that people are once again exposed to such 'hostile environmental conditions'. Threats are environmental and ecological, rather than social or economic. Put another way, once society is removed from view, all that is left are the empirical interconnections between people and their immediate or local environments, hostile or otherwise. With the dissolution of the 'old ways of life' people are exposed to this post-social wild. Within an already-recognizable post-humanist universe of indeterminate multiplicity, the only knowable thing is the observable behaviour of the people involved. Moreover, rather than political change, the only option for improving the survivability of communities existing in the wild is to change or adapt the behaviour of the people who comprise them. Reflecting what David Chandler (2018) has called the ontopolitics of the Anthropocene, in adopting such a stance, we have been suborned into letting the world govern and guide us.

Early warning

Until the beginning of the 1980s, famine was commonly regarded as resulting from externally induced absolute food shortages. The work of Amartya Sen (1981) was important in refocusing famine studies on factors and conditions that were internal to society. Rather than absolute shortages, famines were now the result of the unequal distribution of resources, social capital and market access. Of particular importance was an individual's or group's level of social exclusion. This defined their ability to access that food which was available in the marketplace. During the 1970s, it was still possible to contemplate reducing such inequality through redistributive social democratic measures (Spiz 1978). The enduring importance of Sen's approach, however, was his reframing of the problem in neoliberal terms. That is, by focusing on improving inclusion through strengthening individual capacities and choice-making abilities within a free market (Edkins 2000; Chandler 2013).

During the Sahel-wide famine of the 1970s, the need for a system that could help detect the early onset of famine conditions gained international momentum. Encouraged by academic research showing a correlation between local price fluctuations and anomalous peasant behaviour in times of stress, using aggregated data, the result was the trial establishment of a global early warning system based in the UN's Food and Agricultural Organization (FAO) headquarters in Rome (Cutler 1984a). At the same time, many African governments also established their own national systems (Cutler 1987). Against this background, the fantastic invasion not only gave a fresh urgency to these developments but, given the increased involvement of NGOs working on the ground, expanded the potential range and granularity of the behavioural data that could be collected.

Research conducted by Alex de Waal (1989) on behalf of the Save the Children Fund during the mid-1980s famine in Darfur, Sudan, was influential in shaping the shift towards early warning and the importance of livelihood support. It was discovered, for example, that, due to inefficiencies in the aid operation, humanitarian assistance provided only 12 per cent of the food consumed by the rural drought-affected groups and communities. That the actual mortality rate was lower than this figure might suggest was down to the importance of local understandings and the behavioural changes, or local coping mechanisms, that de Waal was able to describe. Regarding the latter, of particular significance was the triggering of various survival strategies as households adapted to environmental stress.

The alerts included farmers diversifying stock holdings from cattle to more hardy sheep and goats; moving cultivation from rain-land to farms on the banks of seasonal tributaries; livestock and possessions were sold; some household members migrated for work; loans were contracted; and the consumption of gathered wild foods increased. While the idea of 'resilience' would not permeate the social sciences for another couple of decades, de Waal's analysis anticipated its signature affirmation of innovation and adaptation under conditions of stress.

Oxfam's *Roots of Famine* report picks up the interest in early warning. It points out that being alert to the behavioural epiphenomena of distress, such as those pointed out by de Waal, when coupled with nutritional survey data, can alert agencies to the onset of famine conditions. Indeed, even with incomplete knowledge, 'an imminent famine can be seen by watching for key signals' (Cater 1986: 4). In 1986, with funding from several Scandinavian donors and coordinated through the UN system, a two-year project to develop a famine early warning system in Sudan was launched under the auspices of the government's Relief and Rehabilitation Commission (RRC). Besides using government data streams, this project drew on the new information being provided by the field staff of cooperating NGOs like Oxfam and the Save the Children Fund. The data collected included rainfall patterns; planting and harvest outcomes; crop and livestock prices; demographic data, such as population movements and school attendance; and health and nutritional survey information (Eldredge et al. 1986). Having begun the previous year, USAID was also furnishing information on comparative seasonal crop coverage and growth patterns through remote imagery provided by daily satellite overflights (Stover 1987).[17]

Apart from remote satellite sensing, the data collected in the mid-1980s, although mostly gathered by local aid workers, was still a largely direct terrestrial activity. This included the regular completion of market surveys, movement check-lists, individual and group interviews or the physical weighing and measurement of children for purposes of nutritional assessment. This data would then be physically compiled and couriered from the field to Khartoum for manual collation and analysis. Rather than taking days, completion cycles were in terms of weeks, if not months.

As a basic model, early warning is central to the cybernetic *episteme* – that is, behaviour can be converted into signals and alerts. These can be recorded as data and used to both flag anomalies and predict future behaviour. This would develop with growing connectivity to become a core technology with contemporary systems of algorithmic governmentality. Early warning, for example, is intrinsic to a whole range of applications, from cognitive predictive advertising and financial speculation to anti-terrorism surveillance. During its early terrestrial development, however, early warning was inevitably caught up within the ground friction that it generated. Basically, famine early warning has rarely worked (Jaspars 2015). In the 1980s, for example, the growing political push-back by the Sudanese government took the form of restrictions and delays regarding NGO import licences, travel permissions and short-wave radio permits. Friction was also encountered when getting donor and government officials to accept famine as a socio-economic event, rather than just an unmediated outcome of climatological or agricultural perturbation (Cutler 1984b). At the same time, donors were sceptical of government figures, preferring those supplied by the aid agencies. Compounding disagreements on what were the key indicators, crucially, donors were also at odds regarding the exact measures to be undertaken when a probable famine was detected (Cutler 1987).

By the end of the 1980s, the government's donor-funded early warning project within the RRC was already seen as a failure (Eldredge & Rydjeski 1988).[18] Moreover, the NGO experience in Darfur suggested that the rural subsistence economy was no longer viable, having entered a state of permanent emergency. Food insecurity was now seen as a normal condition rather than an exception. Developing de Waal's finding that what people do themselves is more important than the limited humanitarian assistance they receive, disaster management changed to reflect these realities. Since modernist top-down approaches undermine local efforts, the ideal system of disaster management became process-oriented in support of the adaptive livelihood strategies already being pursued by the affected communities (1988). Reflecting an orientation that today's post-humanitarians often consider their own discovery (Meier 2012), the emphasis shifted from emergency relief to a more developmental stance of helping those 'who are already assisting themselves' (Stover 1987: 31). During the latter half of the 1980s, while seldom realized in practice, the ideal or preferred approach to disasters was envisioned as a project-based system of livelihood support (Jaspars 2015).

Conclusion

Intrinsic to the NGO-led fantastic invasion was the transformation of knowledge – which drew on history, society and causal explanation – into information and data that speak to complexity and environment. The disappearance of society, and by implication capitalism, was an intrinsic part of this transition. This ground work would be consolidated with the deepening of connectivity from the end of the 1990s. The fantastic invasion, however, was anticipatory. Unlike the post-humanitarianism that would develop, the ground work involved privileging the collective rather than the personal. Importantly, it also still relied upon presence and human agency. It was also important in helping the spread of capitalism beyond the colonial wage relation into the still autonomous areas of community mutuality and social reproduction. The project form was the organizational means whereby these modalities were first valorized. In examining the livelihood regime, the next chapter considers this aspect of the fantastic invasion.

Chapter 6

LIVELIHOOD REGIME

With the consolidation of the new economy during the 1990s, the idea of the network quickly gained dominance in the global North as a metaphor for the morphological diagram of social life (Castells 1996). The network, with its suggestion of mobility, spatiality and connectedness, was central to the imagining of a post-Fordist spirit of capitalism. While relatively weak compared to the security that welfare-Fordism had offered, the promise was that freedom now lay in this new interconnected, non-hierarchical and meritocratic world. Set against the stifling prospect of a job for life, the new spirit promised creativity and authenticity through the possibility of self-knowledge and personal fulfilment. Rather than of social class, identity was now a matter of choice and self-affirmation. Whether based on location, leisure pursuits, sexual orientation, culture or religion, identity could be multiple and overlapping (Rose 1996). As a signifier, employment or profession came low on the list. During the 1980s and 1990s, when it appeared that capitalism had, indeed, regained some of its former dynamism, any reduction in future security guarantees could be weighed against the 'freedom that has been won, and hopes that the excitement generated by greater autonomy will prove stronger than fears for the morrow' (Boltanski & Chiapello 2005: 93).

In this optimism on the eve of the post-social world, boundaries were transgressed and the dynamism of the network unleashed through the organizational form of the 'project'. Drawing on the management literature from the late 1980s and early 1990s, Boltanski and Chiapello (2005) document how, in an ideal post-Fordist world, people would no longer have jobs for life or a defined career. Instead, the lives of the connected would consist in a series of ever-changing projects. Employment would be a succession of jobs, each of which provides opportunities for appreciation 'and thus the chance of being called upon for some other project'. They call this new urban vision 'the projective city' (2005: 105). Within its environs, as a process of continually connecting and disconnecting monadic existence (Latour et al. 2012), personal freedom translates into a life of changing projects. Individuals meet new people and acquire fresh skills as they pass between different networks. As things change and evolve, some projects and their connections lapse, others are left as a fall-back to be reactivated at a later date. As reflected in this ideal managerial thinking, if, by the 1990s, the spatial metaphor for society was now the network, then, institutionally, the new spirit of capitalism envisaged 'the general organisation of society in project form' (Boltanski & Chiapello 2005: 105).

By the early 1980s, the proposition had already emerged within political ecology that the crisis within welfare-Fordism was pointing towards a world without work. Given this prospect, there was a need to explore a new vision of the future as a series of self-directed activities (Gorz 1982). In testing such a future, Sudan had a limited industrial history[1] and the social state was weak and confined to urban areas. The fantastic invasion encountered a practical opportunity to trial by proxy projectized forms of post-industrial and post-social survival that, due to legacy social democratic constraints, would have been difficult to attempt in the global North. In many parts of Africa, NGOs were able to anticipate livelihood strategies that took for granted the absence of any social responsibility from the state or private sector.

The international aid's livelihood regime (Jaspars 2015: 5–6) can be seen as an experimental projective city in the global South. In all essentials, it no longer exists. By the mid-2000s, it had all but given way to a focus on resilience. As with the previous chapter, here attention is drawn to what was anticipatory about the project form in relation to the cybernetic *episteme*, and where points of rupture exist between then and now. NGOs were early adopters of the project form. During the fantastic invasion, the project was the primary means whereby knowledge was transferred to the community in order to encourage an independent self-acting model of collective development. As with direct humanitarian action, human agency was still important. This time, however, it was vested in the realization of a self-managing community. Such an independent action-oriented approach regarding community no longer exists. As discussed in chapter 11, self-acting technologies are now supported by communities of users permanently enrolled in their endless prototyping. Underlying this rupture, however, there is an anticipatory continuity. This relates to the early understanding of the project as a feedback mechanism for engaging the poor in iterative forms of constructivist learning. This 'learning like a child' approach informed early programming and the design of the human–computer interface.

Project Form

Speaking in the name of people, freedom and rights, during the 1970s and 1980s, the expanding international NGO movement was an important pioneer of postmodernist relations and governance structures. It played a significant role, for example, in criticizing the 'top- down' state-led modernization in the Third World, professional hierarchies and the demobilizing role of the 'expert'. Its voluntaristic ethos echoed the May '68 rejection of the alienation and inauthenticity of consumer society. Organizationally, NGOs were one or two decades ahead of the flat management structures that post-Fordism would come to extol. As an anticipation of the spirit of progressive neoliberalism, Ernst Schumacher's influential *Small is Beautiful: A Study of Economics as if People Mattered* (Schumacher 1974) is important. Schumacher's work represents a rehearsal of the case for NGO direct action through the project form. While dated in terms of language and presentation, his embrace of the cybernetic *episteme* means that his work remains consistent with contemporary cognitive developmentalism (see chapter 12).

In terms of information flow, projects are a longitudinal or iterative series of feedback processes. They impart resources or knowledge to the participant or group while, in return, beneficiaries provide feedback regarding what works or not, thus starting the process all over again. The more projects an NGO manages, the more accumulated information it has; and the more connections and information it has, the more developmental help it can potentially feedback. Writing in the 1960s and 1970s, for Schumacher the state-led development then in vogue benefitted only elites. Its urban and industrial focus excluded the majority of the rural poor. At the same time, it eroded the social fabric of the countryside, leading to uncontrolled urbanization and the harmful proliferation of slums and shanty towns. Apart from wishing to moderate consumer desires among the poor, for Schumacher the aim of development was to get the excluded active, working and, especially, *participating*.

Reflecting then-current developmentalism, Schumacher argued that poverty was non-material in nature. That is, it had more to do with deficiencies in information than with money or resources and, as such, its existence did not warrant radical social change (compare World Bank 2015: 80). Acknowledging the arrival of *Homo inscius*, it lies 'in certain deficiencies in education, organisation, and discipline' (Schumacher 1974: 140). The key ingredient is information. Since the poor only have simple needs, all they require is 'an upgrading by the input of new knowledge' (1974: 167). Based on small catalytic project-based inputs of self-help guidance, simple tools and support, this non-material development 'also has the advantage of being relatively cheap, that is to say, making money go a very long way' (1974: 165). Since getting the poor active and participating was the main thing, for Schumacher it was unnecessary for projects to be economically viable or even of great practical use. Arguably anticipating contemporary precarity, it was more important that everyone was active and doing something, even if creating negligible value, rather than that 'a few people should each produce a great deal' (1974: 145).

In order to administer and oversee the small catalytic inputs required by this information-based developmental model, Schumacher envisaged the formation of international action groups that ideally should be outside of government, 'in other words they should be non-governmental voluntary agencies' (1974: 169). By the mid-1960s, as Schumacher notes, many such NGOs were already involved in this type of grass-roots development work. Prior to the roll-out of structural adjustment in the global South, the main obstacle in the path of this 'bottom-up' NGO development was state-led industrialization and the misguided Third World aim of economic catch-up with the West through centralized planning. The result of the latter was uncontrolled urbanization, increasing poverty and the growth of slums.

Meeting this developmental challenge not only would require NGOs, it would come to nothing unless there was a 'systematic organisation of communications – in other words, unless there is something that might be called an "intellectual infrastructure"' (1974: 169). Through mobilizing administrators, businessmen and communicators, Schumacher envisaged an international *sans-frontières* network of projects interconnecting countries, regions and continents. With international NGOs at the nodal points of such an information network, this matrix would operate as an iterative feedback loop, directing problems encountered in the field to experts who could provide a solution. In an analogue anticipation of cloud computing, it was a system that would not hold information in one centre, but was geared instead 'to hold "information on information" or "know-how on know-how"' (1974: 170).

The 1980s were a period of rapid expansion for the international NGO movement. This coincided with the emergence of network capitalism as a global logistical system and the beginnings of the transition to the new economy of personalized consumption in the global North. While the fantastic invasion played an anticipatory role in this movement, it also reflected its time. There are differences and breaks. Explored more fully in chapter 11, such a rupture occurs in relation to the action-orientation associated with both the 'project' and 'community'. While Schumacher displayed a cybernetic sensibility in relation to the importance of information exchange and feedback, compared to his view of human reason and agency, both would be accorded a more subordinate status following the computational turn. For Schumacher, the best aid was intellectual rather than material, because the former required the effort of understanding and a will to make it work. Without that effort of appropriation, there is no gift. Material goods make people dependent; a gift of knowledge, however, 'makes them free' (1974: 165).

During the fantastic invasion, projects were the means – the pretext even – through which information was passed to communities in order to encourage them to become self-acting collectives. Direct humanitarian action, for example, favoured working through or in collaboration with community organizations or local institutions. Local leaders, elders or dignitaries were sought out by NGOs. If deemed too patriarchal or traditional, new groups were created and women encouraged to join. Realizing the 'community' was a creative enterprise. Such committees and representative groups were thought to provide local aid management with authenticity, especially if the group concerned could be encouraged to target assistance to the most vulnerable – or, at least, to ensure some degree of equity (Jaspars & Shoham 1999). Looking to a future of independent self-development, projects were imagined as building the administrative and management capacity of the participants.[2] The installation of improved village wells or hand pumps was a typical community development project of the period (Redfield 2015). Villagers, for example, would be expected to provide the labour for the installation, together with making arrangements for long-term upkeep or management, in exchange for the NGO-supplied materials. Ideally, such projects should be 'owned' by the community. Ownership in this sense was measured in the extent to which members took collective responsibility for the project aims and outcomes.

This imagined realization of an independent self-acting community, fostered through project work, has little operational meaning for post-humanitarianism. Today, it is objects that are self-acting in the hands of their enrolled user communities. However, the progressive neoliberalism of the fantastic invasion also sought to make markets work for the poor, not by reforming or regulating them, but by providing communities with the knowledge and capacity to realize capitalism's potential. The earlier anti-industrial position of NGOs blurred seamlessly into their 1980s role of preparing communities for the market. More specifically, projects helped to valorize community mutuality and social reproduction preparatory to the incorporation of these formally autonomous areas of subsistence and household production within a socially expansive capitalism.

Social reproduction

The embrace of complexity-thinking and behaviourism undermined modernist notions of relief and development. Initially seen as distinct, relief denoted short-term emergency interventions, while development was a long-term commitment to economic modernization, state capacity-building and catch-up with the West (Rostow 1960). Three decades late, these terms had changed beyond recognition. Relief and development had been internalized within an individual country and, reflecting the 'empiricist onslaught', reduced to poles separated by a shared continuum of beneficiary behaviour (UNDP 1994). Below a certain threshold of behaviour, resources and opportunities, the result is growing food-insecurity and impoverishment. Should this cycle persist, the result is eventual social breakdown. Above this threshold, however, vulnerability is reduced as households embark on a self-reinforcing process of asset accumulation and development (EC 1996). Governors and feedback mechanisms interconnect the different behavioural cycles of decay or growth within their environmental milieu.

In terms of operationalizing the livelihood regime, as a way of promoting positive behavioural cycles, projects were a key form of intervention within the relief-to-development continuum. Varying in scale, the typical NGO project usually focused on the group, community or other local collective. In content, projects ranged widely, from educational interventions to improve health or farming practices, to activities encouraging thrift, prudent investment or, as in well construction or road maintenance, cooperative working. The focus on exclusion and gender was an important dimension of project work. Projects aimed to make markets work by, in effect, preparing or proofing groups and communities for the market. As a means of valorizing community mutuality and social reproduction, projects provided a bridgehead whereby capitalism was able to penetrate the formerly autonomous level of subsistence farming and household production[3] – thus helping to extend capitalism 'all the way down' to the previously autonomous area of social reproduction (Fraser 2012).

Mozambique during the 1990s offers a case-study in this process of developmental inclusion and valorization (Duffield 2007: 82–110).[4] Mozambique had seen its own fantastic invasion during the civil war (1977–92), with the numbers of international NGOs rising from 70 in 1985 to 180 by 1990 (2007: 87). As in Sudan, this was also a time of NGO autonomy, with a similar practice of arm's-length subcontracting between NGOs and donor governments. Following the arrival of peace in 1992, this changed as donors began exerting increasing control over NGOs through closer monitoring and evaluation – in particular, through the introduction of 'log frame analysis' allowing progress to be measured against agreed bench marks (2007: 88–90). During this period, losing their prior independence, NGOs seamlessly morphed into the 'implementing partners' of donor governments. This transition saw a refocusing of NGO projects to target better the poor and the excluded, particularly regarding social and gender issues.

Given the absence of a social state, NGOs act to make communities and groups visible. Through their presence, social surveying and rapid appraisal techniques, they bring communities and groups into the light, making them knowable and thus amenable for governmental interventions (Bryant 2002). Like the mid-1980s famine in Sudan, the civil war in Mozambique was understood as having undermined and dissipated the essential self-reliance of rural communities, exposing them to the wild of an uncertain environment. The view that self-reliance was the natural condition of such communities was axiomatic among aid agencies. In the past, the internationalist New Left had problematized the notion of self-reproduction as encouraging regressive forms of exploitation (Wolpe 1972). Progressive neoliberalism, however, saw things differently.

Providing it could be revived, self-reliance provided an alternative to state-based social protection. Not only was such a protection unlikely to emerge, a resurrected self-reliance would also constitute a free system of social security 'offering possibilities of adaption and strengthening in order to manage the risks of market integration' (Duffield 2007: 93). For NGOs, self-reproduction was synonymous with community mutuality and high levels of social cohesion. Since the civil war was regarded as having significantly weakened traditional relations of self-help, an opportunity existed to resurrect community mutuality through appropriate project work. The intention was to recreate not the old forms of cohesion and reciprocity, however, but a new, more egalitarian system. At a time when the Washington consensus and its focus on economic adjustment was weakening, the renewed focus on the social and internal aspects of livelihood support continued the trend established in the previous decade.

An earlier focus of project work in Mozambique on rich peasants, merchants and commercial groups gave way to the privileging of the marginalized and excluded, especially women, in order to create egalitarian forms of collective empowerment. With gender as a symbol of everything that was backward in traditional society, by the end of the 1990s, rural projects aspired to new forms of social organization and collective ownership through improved self-reliance (2007: 103). Through concepts such as 'structural poverty', the position of excluded groups, such as the elderly or widowed, was seen in relation to a natural economy in which their access to land, labour and communal resources varied over the life-cycle of the household (Fleming & Barnes 1992). The operation of this natural economy compounded the challenge of environmental uncertainty. Domestic responsibilities like birthing and child care, food preparation and agricultural work hampered the market participation of women. Indeed, many women seldom left the homestead (Cuppens 1998: 8–11, 14).

Towards the end of the 1990s, donor-funded projects, including NGO-implemented community road construction, agricultural extension work and land registration projects, were redesigned to include the marginalized, encourage the participation of women and protect against the patriarchy of the traditional system (Duffield 2007: 108–9). The increase in productivity, it was believed, would lift the community as a whole without exacerbating economic differentiation. While largely aspirational, these projects reflected the widespread deployment of the project form as a means of valorizing and integrating social reproduction within a capitalist marketplace. Although the aims were different, in terms of their attempt to create an ideal homoeostatic future, they stand comparison with the colonial practice of native administration, or the creation of artificial systems of 'tribal' authority, even in situations where none had properly existed before (Duffield & Hewitt 2009). In the case of the social project, the vision was one of an idealized, cosmopolitan form of identity-sensitive progressive neoliberalism. This ethos, however, concealed the relations of exploitation inherent to the mobility differential underpinning the project form and the reduction of local knowledge to programmable information demanded by the cybernetic *episteme*.

Programming Poverty

Besides the anticipation of post-social survival strategies and the valorization of social reproduction, the fantastic invasion also introduced another aspect of the project form – namely, its function as a mechanism of exploitation. The project operates as an organizational interface between mobile international aid workers, able to move internationally, and immobile aid beneficiaries. The subsequent appearance of digital humanitarians, operating off-shore through technical volunteer networks (see chapter 10), does not eliminate this spatial effect. On the contrary, it is magnified. Digital humanitarians are even more mobile than their grounded colleagues. As for beneficiaries, not only are they entangled in their local communities, they are contained by an international system that, by the 1980s, had already been closed against autonomous circulation. Enhanced by the asymmetries of information and resources involved, the mobility differential inherent in Schumacher's 'intellectual infrastructure' conveyed a clear spatial advantage to mobile NGOs.

Occupying nodal points in this global intellectual network, NGOs were able to trade their intelligence on local development problems or emergency needs in exchange for government funding, donor patronage or, through appeals and campaigns, public support and influence. The mobility of the exploiter is reflected in the immobility of the exploited (Boltanski & Chiapello 2005: 369–70). Organizationally, since the 1980s, NGOs have kept pace with the marked expansion of the global precariat, growing in numbers and size, widening their mandates and, not least, increasing their budgets. For several decades, it was a successful business model, combining the cosmopolitan high ground with an expanding organizational presence and influence. At the same time, the fantastic invasion launched the careers of innumerable expatriate aid workers who would find themselves climbing the UN ladder, joining government aid departments, going into business or becoming academics.[5]

Local knowledge

Besides being a mechanism of exploitation, the mobility differential also devalues and transforms terrestrial knowledge. The connected world belongs to the mobile, the flexible and the fast. In such a world, there is little room for the detail of area studies, cultural specialism, historical accounts or knowledge patiently acquired through language – that is, knowledge that is entangled in or specific to local conditions. As a result, the whole corpus of academic area studies has been in decline for several decades (Pupavac 2012; Wiley 2012). It is too detailed, hard-won and idiosyncratic for immediate operational purposes. As with data behaviourism, for the practitioner, useful information has to be immediately comparable, non-specific, flexible and capable of being acted upon. Interestingly, however, in relation to indigenous knowledge, an apparent reversal takes place. To the extent that professional or structural knowledge is decried, indigenous knowledge is praised for its practical authenticity. On examination, however, such praise usually reduces to a machine-ready constructivist understanding of the nature of human learning.

When one examines the leading edge of current developmentalism – for example, *Doing Development Differently*[6] or *Thinking and Working Politically*,[7] or the well-received 2015 World Bank report on *Mind, Society and Behaviour* (see chapter 12), one is struck by how little the basic ontological claims made for an empiricist behaviourism have changed since the ground work of the fantastic invasion. Terminology has altered and the presentation may be different but, given the disavowal of theory and radical social change, there aren't that many ways to skin a one-dimensional cat. If you start with a distracted and unthinking *Homo inscius* entangled by feedback loops with its immediate environment – similar to the behaviour of a child – learning is assumed to result from a participatory and iterative process of doing. As a place-holder for machine learning or artificial intelligence, ontologically, developmentalism resembles an automatic vacuum cleaner that eventually cleans a room after repeatedly crashing into every wall and piece of furniture along the way.

An anticipatory version of this is evident, for example, in Michael Edwards' (1989) celebrated piece on 'The Irrelevance of Development Studies'. Drawing on the work of Robert Chambers (1983), Edwards rehearses the already outlined criticism of hierarchical professional and structural knowledge on the grounds of the empirical diversity and complexity of the 'real' world. Development problems are specific to a given time and place. Livelihood and coping systems have evolved over hundreds of years through complex interactions between people and their 'hostile environmental conditions' (Edwards 1989: 120). Given this background, knowledge is not obtained through reason but by trial and error on the part of progressive neoliberalism's *Homo inscius*. The intervention of development professionals short-circuits this process and prevents people thinking for themselves and solving problems their own way: 'development results from a long process of experiment and innovation through which people build up the skills, knowledge and self-confidence necessary to shape their environment in ways which foster progress towards goals such as economic growth, equity in income distribution, and political freedom' (1989: 119–20).

Development is about 'enrichment' and 'empowerment' through the iterative feedback loops made possible by participation. Acknowledging the arrival of *Homo inscius,* this NGO-led participatory process allows for and accommodates the constraints, imperfections, anxieties, fears and emotions that shape the actions 'of real, living people' (1989: 121). Genuine participation, mediated by external but attentive NGOs anxious to 'put the last first', allows these emotions to feed into the learning process.

Computer learning

In the mid-1970s, the architect and computer scientist Nicholas Negroponte, then head of MIT's Architecture Machine Group,[8] was writing about a similar but different problem. To encourage greater public participation, Negroponte was concerned to eliminate the professional from the process of architectural design (Negroponte 2003 [1975]). As suggested, the problem of the 'professional', or rule-based 'technocratic knowledge', was not confined to international development. The autonomy that came from professional knowledge and hierarchical expertise was a target for structural elimination in the transition from Fordism to network capitalism. The move against hierarchy, regulations and the professions was key to the translation of the revolutionary impulse of the May '68 critique into the energy needed to create the computer-based new economy (Turner 2006). This energy, which was vital for tearing down the modernist organizational forms of Fordism, still finds a continuing reverberation in the disruptive strategies of Silicon Valley (Foer 2017; Taplin 2017).

For Negroponte, the issue was not international development, but the nature of human–machine interaction. For him, the elimination of the professional was through computer programming or the creation of 'soft architectural machines', whereby human imperfections, anxieties and emotions could be translated into rich and innovative architectural designs. Removing professionals, preventing them from getting in the way of users and computers, would allow the introduction of 'a simply understood feedback concerning the potential consequences of individual decisions on the whole'. This would help release the 'enormous variety of emotional (intuitive) solutions which can be invented by a large number of future users and might give an incredible richness to this new "redesigned" design process' (Negroponte 2003 [1975]: 359).

Basically, the same constructivist approach of learning by doing that was being advocated by international development was also being applied to understanding human–machine interaction and early computer programming. In the mid-2000s, following the computational turn and the advent of the internet, Negroponte developed the underlying cybernetic *episteme* further in his *One Laptop per Child* programme (Negroponte 2006). Based on corporate sponsorship, the aim of this programme was to make a cheap, robust internet-ready portable computer widely available to children in the global South. Setting aside the organizational details (Negroponte 2007), the theory behind the programme was an inversion of the constructivist programming being explored two decades earlier. The laptop was a means of promoting self-learning among children in the global South through self-directed, iterative user–internet interaction. For Negroponte, this was an extension of the way that children learn naturally through doing: 'we all in this room learnt how to walk, how to talk, not by being taught how to talk, or taught how to walk, but by interacting with the world, by having certain results as a consequence of being able to ask for something, or being able to stand up and reach it' (Negroponte 2006: 1: 53).

The process attenuates when children enter the formal teaching environment of the school. However, 'one of the things in general that computers have provided to learning is that it now includes a kind of learning which is a little bit more like walking and talking, in the sense that a lot of it is driven by the learner himself or herself' (2006: 1: 53).

This idea behind the *One Laptop per Child* initiative was that constructivist computer-based learning would allow beneficiaries to leapfrog the relative absence of a formal educational infrastructure in the global South. With a minimum of instruction, children would teach themselves. While this programme is now defunct, having been outmoded by the rapid spread of broadband and cheap mobile telephony, it is worth commenting on the curious inversion that is evident here.

In the mid-1970s, the problem was one of programming – that is, refining the computer interface to translate human desires, emotions and intuitions into machine code. Three decades later, with the maturing of the internet, the concern was reversed: using a screen interface to subject the internet to an intuitive and constructivist learning process in order to extract topic-specific information and design ideas. Whereas the New Left had hoped to transform the May '68 movement's rejection of alienation and hierarchy into world revolution, the counterculture and progressive neoliberalism have transformed and locked this emancipatory impulse into corporately owned computer technology. This absorption or locking-in is returned to in chapter 12. Computers, machine learning and cyberspace are now the source, originators and custodians of personal freedom and creativity. Rather than overcoming alienation, this is more a redoubling of it. The fantastic invasion's early celebration of constructivist learning among the poor does more than infantilize the precariat. It anticipates that poverty, with increasing connectivity, would itself become programmable.

Conclusion

The livelihood regime has now disappeared and been replaced by that of resilience. Projects still exist but contend with other interventionary technologies and funding mechanisms. Moreover, as we shall see, the action-orientation of 'community' has changed its focus within current forms of technology-based humanitarian innovation. Regarding Sudan, rather than NGOs working themselves out of a job, the fantastic invasion initiated a series of back-to-back emergency operations that stretch from the mid-1980s to the present. Moreover, with the exception of Darfur in the mid-2000s, none of these operations achieved their stated aims (Jaspars 2015: 109). Instead, Sudan – or, rather, the Sudans – has entered a growing list of chronic emergencies. To focus on such issues, however, would miss the point. The ground work accomplished by the fantastic invasion explored in this and the previous chapter, including behaviour-based early warning systems, complexity-thinking, and the valorization of community mutuality and social reproduction, were important in anticipating the post-social new economy then taking shape. These activities were instrumental in the translation of knowledge into behavioural data in advance of the computational turn. At the same time, however, the present represents a break or discontinuity with this period. While extending the cybernetic *episteme*, the fantastic invasion was a period when information was still regarded as empowering self-acting individuals and communities. Since the fantastic invasion, resistance, political push-back and ground friction have increased as connectivity, or the absorption of this autonomy, has deepened. In the face of this resistance, the fantastic invasion is now exhausted and in retreat. The architectural and cultural forms of this retreat are examined in the next chapter.

Chapter 7

INSTILLING REMOTENESS

The post-Cold War attempt to assert a system of cosmopolitan liberal peace now lies buried in the ruins of Iraq and Afghanistan (Richmond 2014). As liberal interventionism unravelled, there has been a resurgence of global resistance and recalcitrance at the same time as a growing appreciation of Western weakness and inhibition (MSR 2017). International professionals who defined themselves as being there on the ground, so to speak – such as aid workers, journalists or academic researchers – are drawing back and circumscribing their presence. Even mass tourism is changing. The once-popular Middle East resorts have emptied, as safer destinations and ocean cruising gain in popularity. Using the example of international professional groups, especially aid workers, this chapter examines this unease. At one level, it reflects the exhaustion of the second wave of terrestrial globalization, of which the NGO-led fantastic invasion was an important part. At another, however, we can see a growing existential remoteness and distance that is proportional to the absorption of human agency and thinking by machines. Erupting onto the international stage in the 1970s, a *sans-frontières* humanitarianism pioneered and developed new forms of connectivity as international space was narrowing and closing against the circulation of political praxis. Reflecting the paradox of connectivity, the international aid industry is now hemmed in by recalcitrance and ground friction regarding the cosmopolitanism it espouses. The transitory architectural form of this exhaustion and remoteness is the fortified aid compound. It is the concrete expression not only of the fantastic invasion in retreat but also of the fact that its heirs now operate from the electronic atmosphere.

Fortified Aid Compound

After an absence of nine years, I returned to Sudan in May 2008. Most of the time was spent in the South helping to complete an evaluation for UNHCR of post-war refugee resettlement (Duffield et al. 2008). Having been in South Sudan on several occasions both before and during the second civil war, which ended in 2005, this trip left a striking impression. With the ending of the war, an increase in the number of aid agencies operating in the South had been expected. What was surprising and unexpected was the widespread withdrawal of donors, UN agencies and the larger international NGOs into visibly fortified aid compounds. The daily routine associated with these compounds was wreathed with security protocols. The situation was extraordinary because international aid workers had enjoyed greater freedom of circulation during the war than, as it then was, three years into the peace.

Mandated to support the peace agreement, the HQ for the UN Mission in Sudan (UNMIS) then lay immediately to the south of Khartoum Airport in a large rectangular compound fortified with double walls and razor-wire, complete with watchtowers and armed guards. Besides a several-storey office block, lines of air-conditioned prefabricated offices within the walls housed the administrative staff. Rows of the UN's ubiquitous white SUVs ringed the defensive perimeter. At first glance, the overtly militarized appearance of the UNMIS HQ seemed out of place. Khartoum remains a relatively safe city and crime levels – especially of violent crime – have historically been low.[1] While UNMIS was there to oversee the peace in South Sudan, it brought the architecture of war into the city. Its HQ, however, appeared anomalous, a sort of mini-Green Zone, but without the obvious dangers or violent history of Baghdad. Given the shameful ineffectiveness of UN peacekeepers to protect anyone but themselves, this was an ironic posture at best. While UNMIS was distinct in terms of the degree of fortification it deemed itself worthy of, since the early 2000s, all the UN agencies and larger NGOs in Khartoum had similarly upgraded their physical defences, variously drawing from the now-common repertoire of no-logoing for buildings and vehicles, erecting outer chain-link fencing and double entrance gates, and employing private security.

Moving on to Juba, the capital of the South, this bunkered architecture was replicated. Having a relatively small built environment during the war, rather than upgrading old buildings, in 2008 Juba had more new, purpose-built aid compounds distinguished by their large spatial footprints. Since the 2005 peace agreement, whole districts had been taken over and divided up between incoming agencies, many relocating from Khartoum and Nairobi. From the outside, the UNDP compound, for example, presented itself as a high white-painted exterior wall, topped by razor-wire. Like an ornamental frieze, 'NO TRESPASSING' was stencilled in blue around its extensive perimeter. The main gate to this complex was complete with a guardhouse, heavy steel doors and crash barriers – entrance by invitation only for ID-carrying visitors.

In South Sudan's small semi-rural towns, the environmental impact of the fortified aid compound was, if anything, greater. In May 2008, UNHCR's purpose-built and security-compliant regional HQ at Yei, near the Uganda border, was still awaiting the completion of some internal landscaping and construction. As a new-build, it brought out clearly the essence of an aid architecture that now seeks to combine defensible working environments with places of cultural refuge. Apart from the usual double gates, guardhouse and outer perimeter chain-link fence, the wide dead zone between the fence and the inner razor-wired wall was patrolled by armed guards at night. This large compound combined accommodation, offices, leisure and essential support facilities. In addition to a water tower and generator making it independent of the town for its basic utilities, the Yei compound contained a block of about a dozen offices and work rooms; a dining area; a laundry; an open-sided tent containing gym equipment; and, in two facing rows separated by open ground, a dozen semi-detached bungalows, each comprising a bedroom, lounge and shower room.

At the time of my visit, the dining area was still a temporary structure fashioned from two metal containers. The intention was to replace these with a brick building and, at the same time, to landscape the rough ground between the facing bungalows. Their freshly painted porches, air conditioning, and internet and satellite TV connections, however, only heightened the feeling of alienation the bungalows gave off – the barred windows and razor-wire on the 8-foot wall behind them unsettling any sense of suburban tranquillity.

Three years later, in August 2011, less than a couple of months after independence, I was in Juba again.[2] As an extract from the situation report suggests, these spacing trends had intensified:

> Juba can best be described as a series of privately guarded gated-communities that provide refuge for its plural elites. These defended spaces vary in their size and degree of autonomy from the rest of the city. In the spaces between these fortified compounds and residential complexes, where the majority of the Sudanese live, there is little in the way of public infrastructure. Having their own generators and guards, and sometimes their own wells, like the agency vehicles that ply between them, these resources are privately owned and managed. Over the past three years the UN has, quite literally, built its walls higher and increased the density of the razor-wire with which it surrounds itself. Contractors, consultants, World Bank officials and international NGO live and work in gated offices and team houses. It is rare to find international NGOs that do not employ a private security company to guard their gates. (Duffield 2011)

One fresh observation yielded by this trip was that the new South Sudan government buildings being erected now mimic the fortified aid compound. While sometimes avoiding razor-wire for more decorative railings to top their high walls, perhaps sensing the catastrophe ahead, South Sudan's elites were also fencing themselves off. Worthy of more consideration than is possible here, this defensive architectural style runs counter to the design of colonial government buildings. Rather than fortified structures, surviving examples and photographic evidence suggest the colonial regime preferred open fronted, low-walled buildings for their administrative centres (Daly & Hogan 2005: 231–52).

Reducing circulation

The fortified aid compound is now widely found, from the Caribbean through Africa to the Balkans, Middle East and East Asia.[3] It is important to emphasize that the bunkerization of the aid industry, and the culture of security that drives it, represents a break with the 1980s direct humanitarian action discussed in chapter 5. Aid compounds, in the modern sense, have existed for decades; they are not new. Regarding the UN in Sudan, these were mostly established by treaty with the government during the 1960s. In these sovereign spaces of the international, such agreements typically confer diplomatic status on international staff and inviolability of the offices, documents and equipment of the agency concerned (UNHCR 1968). Representing visible islands of modernity where vehicles, diesel, electricity, medical supplies and telecommunications are corralled, NGO compounds have existed in South Sudan since the early 1970s (Tvedt 1994). Their present geographical distribution was mainly shaped during the second civil war, being especially associated with the spread of bush airstrips, most of which appeared after 1993.[4] These airstrips provided the in-country logistical grid for the UN's humanitarian Operational Lifeline Sudan (OLS) which, from 1989 to the end of the second civil war in 2005, was managed from Kenya (Duffield et al. 1995: 172–5). Using community labour, the building of bush airstrips was typically agreed with local rebel groups (Levine 1997). As part of these negotiations, rural NGO compounds were established in their wake.

In terms of the current trend to increasing remoteness, the security apparatus that OLS developed and operated during the 1990s is instructive (Karim et al. 1996). Spurred by the killing of four aid workers by the Sudan People's Liberation Army (SPLA) in 1992, this apparatus was designed to move beyond the own-risk ethos of the 1980s, while maintaining humanitarian circulation under the changing conditions of ongoing conflict. In practice, it actually served to increase the number and geographical footprint of international NGOs operating within and moving around the conflict zone. OLS security was based on two interconnected innovations. First was the negotiation and agreement of a series of Ground Rules with the rebel movements, from 1993. These rendered humanitarian assistance dependent on the respect and protection of aid workers and their property (Bradbury et al. 2000). Second, based on the bush airstrips and the establishment of a regional short-wave radio network, came the recruitment of an ex-military security management team that established an intelligence-led hierarchy of security alerts and matched responses. Depending on the threat, these ranged from work-as-normal, through reducing staff numbers, to the evacuation of the location. In the mid-1990s, a review of OLS described this system in the following terms:

> The evacuation and relocation of aid workers, on a few occasions with only minutes to spare, is now a routine event for OLS. As a

consequence, humanitarian assistance closely follows the dynamics of the conflict. This adaptability has increasingly come into its own, for example, as areas of Bahr el-Ghazal and Upper Nile became more insecure from the end of 1994. Here, the system has supported the development of mobile aid teams, enabling workers to remain on the ground for shorter periods, but covering wider areas. (Karim et al. 1996: 2.6.5)

At the time, this security apparatus was praised and enjoyed the confidence of the aid agencies involved. Through informed risk-calculation derived from human intelligence, OLS was able to maintain, even expand, the circulation of international aid workers under conditions of ongoing conflict. This is markedly different from today's more restrictive practices. OLS was disbanded at the end of the war in 2005 and is now largely forgotten. Even before the end, however, not only were aid compounds becoming more defensive in their outward appearance, international aid workers especially were being enveloped in a culture of risk-reduction that works against circulation, exposure and flexibility (da Costa 2012). In 2011, two years before the outbreak of South Sudan's third civil war, there were still a few aid workers in Juba who remembered OLS. The irony of enjoying more freedom under war conditions was not lost upon them.

> During the war I could do all kind of things with UN Security approval that I cannot do now. I was dropped with a radio set and a tent and stayed for weeks in the bush. Today no one walks, no tents. This means that we have no access, nor flexibility to go to the areas where there is greatest need. (Senior UN official, quoted by Collinson & Duffield 2013: 7)

By 2008, compared to the direct action of the fantastic invasion, the interaction between internationals and aid beneficiaries had lost all spontaneity. It typically took the form of a security-approved sortie that minimized time spent outside the compound. These outings were usually set up and mediated by local staff. On the cusp of the rapid diffusion of cheap mobile telephony, even in the rural areas of South Sudan, apart from questionnaire surveys completed by local staff or community organizations, direct information-gathering by internationals still took place through focus group meetings or project workshops. Briefed on the need for gender balance, local staff would go out in advance to organize groups in selected villages or displaced locations. For internationals following behind, UN security protocols dictate that not only is a backup vehicle required, one of their number has to have base radio contact. As in my case, consultants would arrive at an arranged time, conduct stilted and often awkward conversations through interpreters with bemused beneficiaries, before returning to the compound well before dark.

Within a couple of years, even this would change, as the UN began to subcontract its information-gathering requirements to private subcontractors able to operate outside of the UN's own security protocols. Reducing the need for international presence, this reflected a trend that was already evident in the more insecure environments of Iraq, Afghanistan, Darfur, Somalia and northern Uganda – that is, the emergence of a series of techniques known as 'remote management' (Bruderlein & Gassmann 2006; Rogers 2006; Stoddard et al. 2010). This includes practices such as working at arm's length through local staff or local NGOs, or subcontracting information collecting or project management to private companies. In Afghanistan, for example, it was common for international managers, through email or mobile telephone, to oversee projects from a distance – in some cases, never visiting at all (Montgomery 2009: 253). Encouraged by the internet and satellite phone, there are also examples of operational aid agencies no longer being sited in the same country as the programmes they support. For example, Jordan had become a logistical and management base for aid operations in Iraq, while Nairobi and Dubai serve Somalia and Afghanistan, respectively.

The security measures appearing in the mid-2000s can now be seen for what they were: a growing closure to the world at a time when connectivity was deepening. What is important, however, is that this closure and remoteness do not come naturally. They have to be taught, encouraged and rewarded.

Dangerous world

From a conventional perspective, the fortified aid compound is a consequence of the widespread perception that aid work is more dangerous than it used to be. Since the mid-1990s, statistical evidence indicates a steady increase in the number of violent attacks, including kidnappings, directed at aid workers (Stoddard et al. 2009). Typically dominated by a few countries, such as Somalia, South Sudan and

Afghanistan, before declining somewhat in recent years, this violent trend peaked in 2013. With national staff affected more than internationals, there were then 251 incidents affecting 460 aid workers, of which roughly a third were killed, injured or kidnapped (Stoddard et al. 2014). On paper, the trend seems clear; interpreting the figures, however, is less so. Since the 1990s, the total number of aid workers and the insecurity of the countries they are working in have also increased. At the same time, the motives driving these attacks are not always given. Although there have been a number of well-documented political attacks against the UN in the Middle East, what proportion of the remainder relate to criminality or personal grievance is a moot point (Collinson & Duffield 2013).

In accounting for these attacks, since the end of the Cold War there have been regular claims that the nature of war itself has changed. New and often irrational non-state actors have emerged, whose violence shows no restraint and little respect for international laws, agreements or norms (Boutros-Ghali 1995: 42; Kaldor 1999; HERR 2011). Given the horrors of the 'total war' that emerged in the West during the course of the twentieth century, such claims are not particularly helpful. On a firmer historical footing, others have argued that, during the 1990s, international aid mutated with the appearance of the UN integrated mission (Eide et al. 2005). As a key institution of cosmopolitan liberal peace, the aim of the integrated mission was to secure greater coherence between the previously separate aid and political wings of the UN system and their associated NGOs (Macrae & Leader 2000) – as a result, making international aid a key instrument of liberal interventionism (Fox 1999). Geographically widespread, the integrated mission saw the emergence, between the late 1990s and early 2000s, of a number of ambitious terrestrial stabilization programmes that included work on disarmament, demobilization and democratic reconstruction in support of victorious warring parties newly emerging from the long Cold War upswing of internal war.

During the Cold War, development, and sometimes humanitarian assistance, were used to cement international political alliances. Changing the social or political system of a country was the business of clandestine military support to the opponents of the regime – either that, or the political theatre of the *coup d'état*. In comparison, the UN integrated mission was essentially postmodern. The practical fiction of neutrality and impartiality was dropped as aid agencies took sides in support of the victors. Achieving Western foreign policy goals moved from the shadows to become the acknowledged work of aid agencies. Whereas, during the Cold War, NGOs often opposed the international aims of their governments, they were now the 'implementing partners' of the same bodies. Often referred to as the 'politicization' of aid, it reflects our hyper-political times. With the disappearance of 'neutrality', the losers or those otherwise excluded from the peace deal are prone to see a political fix. Consequently, 'the universality of the values promoted by the UN no longer guarantees the security of its access in conflict situations' (Bruderlein & Gassmann 2006: 65). In places like Afghanistan, for example, such politicization led to the effective paralysis of the aid operation during the 2000s (Donini 2009).

While statistics and claims relating to the threats faced by international aid workers are incomplete or ambiguous, one thing is clear. There is a widespread *de facto* perception that aid work – or journalism or academic research, for that matter – is definitely more dangerous than it was. Despite differences on the ground, this belief asserts a structural equivalence between the real dangers of Syria, South Sudan or northern Nigeria, for example, and the more predictable circumstances encountered in most other countries. Although a few war-affected or disrupted countries account for most of the danger, as we shall see, mandatory security and insurance protocols have been centrally rolled out by aid agencies, governments, media organizations and universities, on a blanket rather than a locally nuanced basis.

Importantly, however, reflecting the questionable nature of the structural equivalence being suggested, how individuals comport themselves in a more dangerous world cannot be left up to them. They have to be taught, and correct behaviour rewarded. Now more *Homo inscius* than *Homo economicus,* security cannot be left to the reasoned judgement of the individual or group. The World Bank, for example, in its 2015 report on *Mind, Society and Behaviour*, has a section on the cognitive biases and confusions that challenge aid professionals and can impair their judgement (World Bank 2015: 179–91). In relation to security, people have to be instructed in how to read an environment from a post-human perspective. Since the world lacks history or causation, the task is to experience the enfolding environment as a shifting sea of unmediated cognitive cues, signals and alerts. In response to uncertainty, international aid workers, for example, are trained not only to scan the horizon constantly, but to accept defensive living and endless risk assessment as good for themselves, their wellbeing and their work. Given the gaps and ambiguities in the evidence, however, the training regime, with its associated insurance, career and security requirements, is more real than the dangerous world it defends against.

Securing failure

By the mid-1990s, the UN's first post-Cold War, system-wide attempts to work in situations of ongoing conflict in places like Somalia, Bosnia and Rwanda were breaking down.[5] Against this background of failure, the need for better security training among international aid workers first emerged (Cutts & Dingle 1995). From this time, improving security training for aid workers, especially enhancing risk-perception, has been an ongoing issue (Van Brabant 1998). A key publication was Koenraad Van Brabant's (2000) *Operational Security Management in Violent Environments*. Drawing from earlier ad hoc NGO programmes and training initiatives, *Operational Security* brought together in a comprehensive manner what has since become a standard training template (compare RedR 2017). It is the nature of security training to encourage standardization – having different organizations doing different things is counterproductive and harms interoperability. Security encourages the centralization of decision-making within organizations. In the case of aid agencies, for example, it has justified transferring much day-to-day managerial responsibility from field offices to HQ managers. A good example of this centralizing tendency is the 'kill lists' of suspected terrorists operated by the Obama administration. At the apex of the 'kill chain', the former President himself was responsible for the final life-or-death decision (Ackerman 2015; see also Chamayou 2015 [2013]).

Regarding the security of international aid workers, the generic training framework that has emerged typically divides field security into a number of scenarios, including that of outside movements, the work and home environments, and personal wellbeing. Training programmes exist in basic or advanced forms, they can last from several hours to several days, and vary in realism from classroom examples to outdoor role-play exercises, including car-jacking and hostage-taking. In its essentials, the development of field-security training within the UN builds upon and consolidates earlier NGO initiatives. The security training done by the UN is important because it sets the standard and acts as a point of reference for donors and other aid agencies. As a means of risk reduction, donor decisions on whether to fund a particular NGO, for example, can take into consideration the degree to which the organization is compliant with UN security practice.

While individual UN agencies began by developing their own policies, the trend has been towards a 'system-based security approach' (Bruderlein & Gassmann 2006: 65), resulting in centralization and standardization. This occurred at the same time as the already-mentioned appearance of the UN integrated mission and the politicization of aid. A system approach emerged out of increasing cooperation between the Department of Peace-keeping Operations (DPKO) and the Office of the United Nations Security Coordinator (UNSECOORD). This cooperation focused on creating uniform security standards and procedures, including comprehensive security and stress management training (UN 2001: 3–4). By 2002, complementing individual agency initiatives, the UN was conducting one-off security training sessions in more than 100 countries.

At the same time, Minimum Operational Security Standards (MOSS) were introduced. MOSS represents the development of an objective set of security standards covering security planning, training, communications and equipment, for implementation at each UN duty station. These minimum standards spell out 'the standard which must be met in order for the system to operate safely' (UN 2001: 6). Importantly, the adoption of MOSS, and more recently the Minimum Operational Residential Security Standards (MORSS), also became an essential requirement of the UN's insurance underwriters. While MOSS/MORSS standards can vary, they constitute a set of centrally driven minimum operational requirements. In practice, this means that all UN operations have to be MOSS/MORSS-compliant. Propagated in the politicized medium of the UN integrated mission, fed by insurance requirements and driven by security experts, such developments shape the conformity culture of the fortified aid compound.

The August 2003 bombings of the UN and International Committee of the Red Cross (ICRC) HQs in Baghdad added further impetus to the emergence of a centralized system-based security apparatus

(Bruderlein & Gassmann 2006). Headquarters oversight was strengthened and, following an improvement in the career prospects of security personnel, standardized security protocols were rolled-out through what was emerging as a global network of security officers. In December 2004, a new UN Department of Safety and Security (UNDSS) was established within the UN Secretariat. This brought together existing security personnel, such as UNSECOORD and the civilian security components of DPKO, under one roof. In January 2005, a former Assistant Commissioner of Scotland Yard was appointed to head DSS at Under Secretary General level. With a specific mandate 'to professionalise the UN security system' (2006: 76), this was the first time a security professional had been appointed at such a senior level within the organization. One outcome of these developments is that standardized security training is now mandatory for all UN staff.

This security apparatus came together in the mid-2000s, at a time when the depth of the foreign policy debacle in Iraq and Afghanistan was coming to light. It emerged into adulthood as liberal interventionism and the UN integrated mission were being buried in the rubble of the Middle East. Henceforth, interventionism would take a back seat. It would give way to demands for resilience as the inhibition resulting from fears of terrestrial blow-back took hold (Chandler 2016b; Joseph 2016). As Peter Sloterdijk pointed out, those who would intervene to police or democratize have been forced to acknowledge 'that all initiatives are subject to the principle of reciprocity, and most offensives are connected back to the source after a certain processing time' (Sloterdijk 2013 [2005]: 11). In this emerging climate of failure and inhibition, the mid-2000s also marked a rapid increase in connectivity, with the availability of broadband and the accelerating penetration of cheap mobile telephony within the global South. Future-present imaginaries of a technological solution would henceforth qualify failure and inhibition. If the ground was now corrugated with friction and resistance, the electronic atmosphere could provide a fresh vantage point. The question arises, however: how do you now understand a distant and recalcitrant world and, not least, act within it? As a new world of political push-back, unknown threats and international no-go areas? In the case of aid workers, shaping this understanding and comportment has been the central task of field-security training.

Personalizing Security

In the mid-1990s, prior to the emergence of an expert UN system-based security regime, we have already seen that OLS was operating a well-regarded field-based security system in South Sudan. Despite them being separated by only a decade, the differences between the two approaches are telling. Managed from northern Kenya, OLS' security system was locally embedded and responsive to the dynamics of the war. Aid workers, including internationals, moved in and out of South Sudan according to the changing rhythm of the conflict. System-based security is different. It changes the basis of security and redirects its focus. Whereas OLS allowed space to understand the local dynamics of conflict through grounded informants, situation reports and causal reasoning, there is no longer an interest in such detail. As environments are not as open to the immersion and ethnographic investigation of the past, they are evacuated of history, politics and causation. The resulting 'complexity' is normalized by encouraging individuals to adopt a direct or unmediated cognitive relationship with the outside world while, at the same time, linking this relationship to the health and fitness of the inner self. System-based security teaches the individual how to read and understand the environment, *any environment*, as a changing mix of green, amber and red behavioural signs and alerts, and how this ability is enhanced through strengthening one's personal mental resilience.

Emerging towards the end of the 1990s and early 2000s, the privileging of cognitive immediacy finds parallels and structural echoes across a whole range of practitioner, educational and aesthetic fields. The individuation of environmental experience underpins, for example, current trends in architecture (Spencer 2016a), as well as the rise of post-humanism within the academy (Braidotti 2013). Alison Howell (2012), for example, demonstrates how an earlier discourse around post-traumatic stress disorder (PTSD) as a mass medicalized condition justifying liberal interventionism (Summerfield 1996; Pupavac 2001) has given way to encouraging personal resilience combined with neuro-cognitive approaches to work-related stress. Field-security training based on environmental awareness and care of the self creates portable forms of transnational expertise that aid workers carry from one 'challenging environment' to another. In relation to the international, one is tempted to draw the conclusion that, as a consequence of the failure of liberal interventionism (Richmond 2014), the West has given up trying to understand the world. While one could try to rekindle an interest or raise the alarm, there is the suspicion of a certain political functionality in this. With global precarity growing and, importantly, blurring North–South divisions, blocking off the world as dangerous and unknowable can have its uses.

Permanent threat

As a means of normalizing remoteness, security training has a number of generic characteristics. While the evidence on the dangers facing aid workers contains unknowns and ambiguity, security training strips out all shades of grey. It adopts an uncompromising view of the external environment.

> *Internationals everywhere and at all times face permanent and pervasive danger.*

Such unequivocal sense-certainty is understandable. The purpose of security training is to encourage behavioural change in order to strengthen organizational, individual and mental resilience. It cannot do this if the message is overlain with the noise of doubts or exceptions. At the same time, the clarity of the message means that training materials lend themselves to deconstruction. They cast light onto remoteness and environmental individuation more generally.

In Sudan in 2008, it was necessary for visiting HQ staff or consultants to pass the UN's Basic and Advanced Security in the Field training modules (UNBSF 2003; UNASF 2006). Without passing these modules, you could not get a UN ID card and therefore enter UN compounds, board UN flights or travel in UN vehicles. You were condemned to exposure outside the archipelago of defended international space. The training modules came in two interactive CD-ROMs combining voice-overs, video clips and role-play exercises with multiple-choice end-of-level tests. The Basic and Advanced modules both culminate in a final examination. The identity of the trainee is password-protected and an animation along the bottom of the computer screen records progress through the various levels. Reflecting the tone of the training, the animation features a white UN SUV travelling along a twisting road bordered in places by trees that might conceal a threat. Correct answers incrementally advance your journey to the safety of the next gated-complex. Get it wrong and your SUV is knocked back, remaining longer in no man's land. Each CD takes about an hour to work through. Upon successful completion, the software prints a named pass certificate.

Some questions provoked amusement among the aid workers with whom the training was discussed. In general, however, they enjoyed the experience and appreciated the advice given. It provided assurance that the UN was serious regarding duty of care. While there is an established practitioner literature on field security in hazardous environments (Van Brabant 2000; ECHO 2004; Stoddard et al. 2006; Stoddard et al. 2010), this work is part of the security apparatus itself. It does not problematize its subject or analyse its consequences. On the contrary, training is taken as a self-evident good. In wishing to step back to gain a more critical perspective, the point of contention is not whether a given situation is dangerous or safe, or if the advice given is good or bad. Something far more fundamental than these empirical questions is unfolding.

A liberal apparatus of security has emerged that has replaced an earlier orientation to risk based on the accepted existence of *Homo economicus*. The rational subject was trusted to make considered decisions on the basis of available evidence (Pupavac 2001). This was the culture current in the 1960s and 1970s, when the international was a space of political possibility. Security training exemplifies the institutionalization and governmentalization of risk management to produce conformist and inhibited subjects. The UN's security training is not optional but a mandatory MOSS/MORSS requirement. Apart from access to the UN system itself, it is essential for claims under the UN's Malicious Acts Insurance policy. Any loss or injuries suffered in breach of security directives are void. Coupled with aid agency fears of litigation claims alleging poor duty of care (Butler 2003), a powerful governmental apparatus for reducing autonomy and changing behaviour has emerged. Although limited to the global North–South interface, field-security training is an example of a far more general and widespread governmental apparatus.

Structured around the above prime message, all of the UN's required behavioural change can be derived from it. In its opening section, the Basic Security CD-ROM quotes Mary Robinson, the former High Commissioner of Human Rights, to underline that 'some barrier has been broken and anyone can be regarded as a target, even those bringing food to the hungry and medical care to the wounded' (UNBSF 2003: Module 1: 2). In different ways and contexts, the prime message is repeated throughout the modules. Security training reinforces the idea that times have changed. Like it or not, aid workers now face pervasive threats from a calculating and unpredictable enemy. Since this enemy is faceless, follows no particular pattern and can strike anywhere at any time, security demands constant vigilance regarding the external environment. This relentless pressure also draws attention to one's own inner vulnerability and psychological strength. The onus is on the resilient aid worker to demonstrate responsibility by making the right choices and thinking the right thing: 'In certain countries, the advice will be to stop when your vehicle runs somebody over on the road; in another setting, the advice will be certainly not to stop until the next police post' (Van Brabant 1999: 9).

Training imparts a particular way of experiencing and acting within a given environment. It covers things like travelling outside the work place, home and office security and personal safety, including how to respond under fire, or if hijacked or taken hostage. Regarding road travel, for example, things like check-point etiquette are rehearsed, together with how to behave with child soldiers and react to weapons. Reading the road is important, like slowing down on the approach to traffic lights in the hope of avoiding stopping. In selecting a home neighbourhood, the aid worker should look for positive environmental cues like the level of street lighting, numbers of pedestrians, traffic volume and parking facilities. Urban segregation helps because families 'with similar income levels tend to share similar lifestyles and security concerns' (UNBFS 2003: Module 2: 6). Inside the home, attention is drawn to locks, window bars and alarms. Similarly, in the office, things like a secure reception area for screening visitors, using the front desk as a defensive structure, having barriers in interview rooms and a secure office bolt-hole are recommended. Advice is also given on how to defuse tension and handle hostile crowds.

Working through this training involves completing numerous small tests and end-of-section exercises in order to get to the next level. For example, the actions that aid workers should take when first arriving at their duty station are rehearsed as a yes–no exercise. Questions requiring 'yes' include: do you seek a security briefing, meet your local warden, register your family members with the office, and enquire about medical services? In contrast, the 'no' questions are: do you 'check the area around the office and your residential areas on foot?', and 'try food from local food vendors?' (UNBFS 2003: Module 2: 16). In a similar fashion, the training module CDs go through a range of different scenarios and threat environments. In order to move to the next level, and complete the movement of your SUV quickly across the bottom of the computer screen, the desired responses reproduce the behaviour of a responsibilized, risk-averse, segregated and inward-looking subject.

Internal strength

While vigilance and risk management allow humanitarian rescue to be tasked, an aid worker's prolonged exposure to suffering and uncertainty also creates cumulative levels of stress that undermine mental fitness and thus personal effectiveness (Blanchetiere 2006; Comoretto et al. 2011). Tackling this 'vicious' feedback loop is an important part of an organization's duty of care and an expanding area of resilience training. Inner resilience is important. Indeed, in the absence of an outside world structured by history, politics and causation, strong mental resilience among individual workers is essential for the successful completion of an organization's humanitarian mission. In a complex and uncertain environment, taking responsibility to nurture and strengthen one's inner self is a determining factor. Having origins in the military, resilience training has spread across a range of practitioner and educational activities, including the humanitarian field (O'Malley 2010; Howell 2012). As a practice, the centrality of mental resilience privileges the fortified aid compound as a necessary physical refuge from outside uncertainty. Its walls and razor-wire, together with its enhanced connectivity, enclose a safe therapeutic space.

The importance of nurturing the inner self for coping with stressful environments has been intrinsic to field-security training since its early days (Van Brabant 1998). Prolonged exposure to the trauma of violence and loss can produce cumulative neuropsychological effects such as anxiety, flashbacks and depression, together with excessive alcohol consumption or similar risky behaviour, among aid workers. While most of the literature focuses on direct exposure, as an example of the social 'generosity' of psychology (Howell 2012), similar stress effects have also been claimed by remote technical volunteers, such as crisis mappers working on digital data (Jarmolowski 2012; Meier 2015). Stress effects are not just physiological, they are also neurological. Apparently, the brain itself physically changes as a result of prolonged extreme stress. Certain structures (the hippocampus and amigdala) are altered, resulting in the brain's reduced ability to buffer shocks, while inappropriately triggering trauma responses – for example, at the sound of a car backfiring (IRIN 2010). We are unable to examine critically such claims here (see Stadler 2014). It is possible, however, to see how such a powerful, medicalized imaginary helps to shape behaviour and structure space.

In order to reduce negative stress, the common recommendation is to make life as 'normal' as possible (IRIN 2010). While organizations can do things like providing training, counselling and improving living conditions, much of this is down to individuals. What we are seeing in an anticipatory form is personalized biopolitics of the biohuman. Healthy eating, physical exercise and regular sleep while avoiding excessive, reclusive or risky behaviour. Having recreational or fun activities to pursue during down-time, together with arranging to personal taste the furniture and fittings of one's immediate life, so to speak, is important (Achilles Initiative 2013). Work also needs to double as a social support system. A buddy system, for example, is important for the recognition of burn-out in oneself and others, as well as providing empathy and sympathetic support. Agencies should encourage and make it possible for aid workers to keep in touch with distant family and friends through email and social media (IRIN 2010).

The fortified aid compound is not simply a defensive structure against outside threat and uncertainty, it provides a physical refuge where regular living and positive therapeutic inner strengthening can be practised. The need for a barrier against the outside, and the down-time it affords, are also reinforced by another consideration. While PTSD discourse has now waned, one aspect of its medicalization of stress lingers on. Like an infectious disease, unless precautions are taken, the trauma of disaster victims can pass to aid workers. Having a humanitarian empathy for victims 'does not mean sharing their diseases. To work effectively for others you need to be healthy' (Dr Gro Harlem, former Director General, WHO, UNBFS 2003: Module 5: 2). The contagion of external suffering adds urgency to the need for a refuge where aid workers can switch off.

Resilience training goes further than supporting regular living and keeping in touch. Through the notion of *positive thinking*, it encourages the active suppression of negative thoughts and memories regarding the outside world (Achilles Initiative 2013). Besides the importance of meditation and relaxation techniques, this involves an emotional acceptance that some things cannot be changed. In addition to eating properly, exercising and hanging out with friends, positive thinking requires taking time to refocus on affirmative thoughts, small wins and good outcomes. Positive thinking is not simply reactive. Done properly, the aid worker can emerge personally stronger and more confident (IRIN 2010). In line with resilience thinking more generally, out of trauma and stress something new and better can emerge (Folke 2006). From disaster, you can 'bounce back better' (DFID 2011). What's needed is for the individual to find out what works best for them and to practise it.

Emphasis on the inner self, and the need to protect against neurological brain damage, reinforces avoidance and remoteness from the world. They normalize segregation and reliance on increasing connectivity to bridge the distances and jump the spaces. While the fortified aid compound is shaped by security protocols and insurance requirements, it also answers inner needs and personal vulnerabilities. It provides a secure space for the governance of the self amid surrounding complexity. However, given what has been said about trauma and inner vulnerability, we are also faced with the dilemma that humans cannot be entirely trusted with the job. Humans can never be completely relied upon because, to a lesser or greater degree, they may be burnt-out or brain-damaged. Like many others in contemporary society, aid workers have come to self-identify with neoliberalism's *Homo inscius*. This identification helps explain the now fashionable trend to include aid professionals themselves, not just beneficiaries, within any discussion of the cognitive constraints and challenges facing development (World Bank 2015). In the last analysis, the fortified aid compound is an architecture not of expansion but of retreat. As connectivity increases, together with the costs, recalcitrance and friction associated with ground presence, the fortified aid compound could well be a transitory phenomenon. The more people retreat from the world, the fortified compound, despite its protected layers, can start to look too close to danger and uncertainty.

Conclusion

By the late 2000s, not only was the widespread retreat of the 'international community' into fortified aid compounds and gated-complexes remarkable, the indifference of the agencies to the incongruities of the built environment they were actively creating was also notable (Smirl 2008). Defensive and segregated living had been normalized. Rather than evoking any sense of alienation or lost perspective, UNHCR's Yei compound in South Sudan, for example, was typically admired for the superiority of its therapeutic 'deep field' facilities. Already regarded as a hardship posting, one reason for building such well-specified secure compounds was because experienced international staff would not otherwise come to South Sudan. As a node in a networked archipelago of international space (Petti 2007), the fortified aid compound is more than a defensive structure. It is an extension of inward-looking Western therapeutic culture into the post-colony.

As an object of architectural design, the outward banality of the fortified aid compound has little in common with the warped and fluid parametric architecture that is now appearing in metropolitan shopping malls, science centres and manufacturing parks. Douglas Spencer (Spencer 2016a: 65) has shown that such flowing and curvilinear design seeks to capture and valorize 'an underlying, and essentially emancipatory, order of complexity'. While complexity is embraced in the metropole, and celebrated in the sensuous and unmediated exteriority of things, in the post-colony, in the wake of the failure of liberal peace, the banality of the fortified aid compound suggests that what is being valorized is inside. As a nodal point in a shrinking archipelago of liberal international space, it is the interior in the broadest sense that is valued – not just the enclosed support and leisure facilities but, more importantly, the bunkered therapeutic space that allows for a practised mental disconnection from the complexity of the traumas, threats and uncertainties circulating outside.

At the time of writing, *To Kill a Mockingbird*, Harper Lee's classic novel about racism in the American south, had been pulled from the reading list in a Mississippi school because the language used 'makes people uncomfortable' (Agencies Mississippi 2017). This is one example of a growing trend for students in the USA and UK to demand campus censorship and 'no-platforming' for ideas or people who are deemed inappropriate or personally unsettling (Haidt & Haslam 2016). During the late 1960s, the university campus was envisioned as a potential liberated space where radical students could reach out to other radical forces in the interests of world revolution (Barnett 1968). Today, the university campus is more like a fortified aid compound – that is, a place of refuge where you can seal off the world and protect the inner self.

In 2010, when liberal interventionism was in retreat, I wrote that the fortified aid compound was symptomatic 'of the aid industry reaching a strategic dead-end' (Duffield 2010: 471). Today, this needs qualification. What had drawn to a close was the second wave of terrestrial globalization, of which the fantastic invasion and liberal interventionism were part. In retrospect, the development–security nexus (Duffield 2001) which provided a framework for analysing this expansion was, essentially, a ground-based apparatus. In a bid for global governance, it addressed terrestrial forces and physically present agencies. It was concerned with orchestrating direct relationships and marshalling material resources. Rebuffed and pushed back, the dead-end in question was this nexus of *terrestrial* forces. What I failed to realize was that this apparent impasse was in fact a point of departure. Following the computational turn, and the rapid spread of cheap mobile telephony throughout the global South, the cultural dead-end of the gated-complex marked a liberating and celebratory leap into the verticality of the electronic atmosphere (Elden 2013); the last unregulated global plane where capital accumulation, acts of piracy and one-sided political violence are still possible (Weizman 2002; Chamayou 2015 [2013]). The fortified aid compound, or at least the remoteness, inhibitions and mental vulnerabilities it embodies – the world alienation, if you will – signals that the present third wave of hyper-bunkered globalization is now well under way. The next chapter begins to address this new strategic plane.

Chapter 8

EDGE OF CATASTROPHE

Around the mid-2000s, a tipping point was reached within the international aid system. The relations, expectations and modes of governance that had taken shape some two or three decades earlier during the NGO-led fantastic invasion (see chapter 5) were finally retired. Rather than community-based self-reliance and direct forms of humanitarian intervention, ideas of resilience and practices of remote management were gaining ground. Spurred by austerity, policy failure, political push-back and growing risk aversion, the aid industry was ready for renewal and fast opening to private-sector funding, commercial sponsorship and the promise of information technology (Zyck & Kent 2014). Regarding this period of change, it is useful to speculate on what the rise of resilience thinking tells us about the state of capitalism. Resilience suggests that, providing system functionality can be retained, it is normal for systems to suffer perturbations and periodic shocks that propel them from one condition of temporary equilibrium to another (Holling 1973). Whatever this may tell us about ecology, it is an apt description of the last three decades of network capitalism's periodic crises and poor economic performance (Streeck 2011). Rather than expanding, capital is now challenged to find ways of working with and through the modalities of permanent emergency – and, if it can, to capitalize on crisis conditions (Klein 2007).

Gathering speed in the 1980s, the world of work in the global South has become increasingly casualized and deregulated. In terms of its political significance, one of the main achievements of neoliberalism has been to create an ever-expanding global precariat (LeBaron

& Ayers 2013). While casualization is now gaining ground in the North, by the end of the 2,000s, a clear majority of Africa's non-agricultural workforce was engaged in the informal or shadow economy (Meagher 2016). The unstoppable growth of the precariat has authored a major policy change towards the informal sector. Rather than ignoring or fearing informality or, at best, wishing to graduate it to the conventional economy, the aim is now to incorporate its extended shadow networks directly into global business chains (2016). Not only does the 'bottom of the pyramid' (Prahalad 2006) represent a huge potential labour and consumer market, but also of crucial importance to the attraction of the informal sector is that, perforce, it comes complete with its own shadow means of social reproduction. For a free-rider capitalism that no longer creates permanent jobs or provides social protection, the latter is a valuable source of self-reproducing or autopoietic vitality. By its nature, however, the precariat exists in a dynamic relationship between economy and disaster. Made possible by the spread of mobile telephony and data informatics, the contemporary boomerang effect inhabits and explores this relationship. Earlier chapters examined humanitarian disaster as the site for the preparatory ground work in translating knowledge into data. Looking to the future of capitalism, the global South still functions as a laboratory – this time, for the disaggregated biopolitics of permanent emergency that seeks to secure for capital the autopoietic qualities of precarious life.

Global Precarity

Precarity denotes a condition in which the casualization, informalization or unpredictability of work coexists with economic vulnerability, environmental uncertainty and an openness to surprise and shocks. Borrowing from ecology, one could say that the precariat live on the edge of catastrophe. They are constantly engaged in a daily struggle to avoid the probability of extinction. While for decades a condition of the global South (Munck 2013), to the extent that precarity is now being embraced as a viable post-social global future, this is a new historic dispensation. As a lived experience, the idea of precarity performs a valuable refocusing role. It shifts attention, for example, from statistical measures of poverty to the action-oriented ethnography of how people survive in conditions of permanent emergency. While a shallow optimism can be derived from contingent declines in estimated levels of 'extreme', or $1.90-a-day, global poverty (Collins 2016; Kristof 2017), we have to ask whether the extent to which people are above or below such arbitrary lines is less important than the constant uncertainty and everyday hassle that engulfs them and thereby constantly threatens their social reproduction.

Social reproduction encompasses the need for humans to reproduce as a biological species. Historically defined, it includes birthing and caring for the young, sick and old while maintaining family, friendship and wider community linkages, identities and moralities (Fraser 2016). Without these often unremarked but vital functions, traditionally unpaid and cast as women's work – although men have always done some – capitalism would not exist. Now unrestrained, the new economy's conscious aim to drive down wages, extend hours of work and remove social protection has not only exacerbated the crisis of social reproduction, it has revealed afresh capitalism's dangerous propensity to destroy and consume its own conditions of existence (Crary 2014). This crisis, however, is coterminous with the relentless penetration 'of media technologies into life through a frenzy to record and store information' (Halpern 2014b: 224). For many, the advent of data informatics is providential regarding current complexities and logistical challenges. Rather than tackling root problems, however, data informatics harnessed to resilience thinking attempts to maintain the status quo by avoiding or working around obstacles and ground friction. A post-humanitarianism has emerged that, while constantly having to address the consequences of precarity, nonetheless sustains it by dint of trying to make its incumbents resilient to the radical uncertainties of their existence (DFID 2011; Betts & Bloom 2014; Meier 2015).

By no means limited to urbanization, precarity in the global South is associated with the rapid post-colonial growth of slums (Davis 2006). Gaining momentum with the easing of colonial residence restrictions, socialist-inspired economic catch-up and 'drain the swamp' rural counterinsurgency techniques, the explosion of urban precarity has largely been the result of economic liberalization (UN-Habitat 2003: 45–6). Austerity in the modern sense was first imposed on the global South at the end of the 1970s by the International Monetary Fund and the World Bank in the form of 'structural adjustment'. Comprising a mixture of enforced expenditure cuts, economic deregulation, downsizing of public bureaucracies, infrastructure sell-offs and cuts to subsidies in exchange for state debt refinancing, austerity was introduced aggressively in contexts that lacked the cushioning mechanisms that then existed in the North. At the time, it was criticized for its disastrous impact on public welfare (Cornia 1987; Walton & Seddon 1994). Less commented upon, however, was the fact that imposed austerity also encouraged a major expansion of what, in the global South, is variously called the informal sector, the black market or the shadow economy. Labour and entrepreneurial activity were forced out of the formal sector and the household and, having nowhere else to go, into the shadow economy (Duffield 2001).[1]

Registers of informality

The notion of informality captures a raft of production, trading and service activities that, together with the employment opportunities created, operate outside state control and formal tax and regulatory requirements. Lacking legal recognition and statutory protection, workers in the informal sector operate below the radar of economic conventionality (Nordstrom 2000; Roitman 2001). The 1980s saw the expansion, for example, of extensive local–global transborder trade networks across Africa, dealing in all manner of foodstuffs, manufactured items and consumer goods. These extra-legal conveyor belts provided employment and commodity access for millions of men and women (Meagher 1998). At the time, estimates of the size of the shadow economy across much of the global South were as high as 40 to 50 per cent of official GNP (Duffield 2001: 142). Earlier views of the informal sector tended to see it as the epitome of underdevelopment, and hence a transitory phenomenon that was destined to disappear with development. The global trend for casualization and shadow networking, however, has been consistently upward (Jutting & Laiglesia 2009). At the turn of the twenty-first century, after decades of failed NGO and World Bank remedial interventions, the shock of the 'astonishing size of informal economies' was initially greeted with alarm (Meagher 2016: 485). UN-Habitat's unflinching 2003 Global Report on Human Settlements, for example, provided much of the empirical evidence for Mike Davis' aptly entitled exposé *Planet of Slums*, published in 2006.

With around a quarter of the world's population affected, postcolonial urbanization has largely been an exercise in the relentless expansion of urban informality (WEF 2016). In Africa, for example, the number of slum dwellers more than doubled between 1990 and 2014, to around 201 million or over half of the continent's total urban population (UN-Habitat 2016: 84). While slums are only a rough proxy for precarity, they constitute 'off-grid' milieus. Usually situated on marginal land, often low-lying, on slopes or contaminated industrial sites, slum dwellers commonly endure overcrowding, inadequate shelter and insecure tenure coupled with an absence of regular electricity, water or sanitation services, and, at most, basic medical and educational provision (UN-Habitat 2003: 10–12). Given added impetus by the 2008 financial crisis, by the end of the decade, depending on location, estimates of between 70 per cent and 90 per cent of Africa's non-agricultural labour force, more than half of which is comprised of women, were now in casualized, low-paid vulnerable employment. Moreover, as the OECD declared, by this time most of the 1.7 billion people then defined as extremely poor 'depend on their labour for survival *as it is often their only asset*' (emphasis added, Jutting and Laiglesia 2009: 19). Based on the UN threshold of $1.90 a day, this suggests that a billion or so dispossessed and otherwise assetless people currently earn less than $700 a year. To get a feel for the ball park we are in, this amounts to around $21,000 over a lifetime of active work.

It is important to reach behind such abstract statistics to try to grasp the dynamic at work. Since the 1980s, the global South has been trapped in a process of 'jobless' growth, in the sense that the creation of formal employment has either ceased or, relative to demand, is now insignificant. What is new is the dissolving of subsistence and semi-subsistence economies without, contrary to the experience of nineteenth-century Europe, the appearance of a formal alternative. Instead, there is precarity that is, perhaps, best described oxymoronically as *active unemployment.* As a global phenomenon, informalization and the crisis of social reproduction are reflected in the unprecedented growth in social and economic inequality. Wealth has been transferred globally from the masses to the elites between and within both rich and poor countries (OECD 2008; Piketty 2014). According to Oxfam, the richest 1 per cent have now accumulated more wealth than the rest of the world put together (Oxfam 2016). At the same time, 'the wealth owned by the bottom half of humanity has fallen by a trillion dollars in the past five years' (2016: 1).

Much has been made of the statistical decline in extreme global poverty since 1990. However, as argued in chapter 11, the resilience regime has changed what counts as evidence, preferring a narrow empiricism that black-boxes wider social and political considerations. At the same time, apart from being uneven,[2] the decrease tends to occlude an important growth in precarity. In 2013, some 767 million people, around 11 per cent of the world's population, were said to be living in extreme poverty, a drop of around two-thirds compared to the 1990 estimate (World Bank 2016b: 4). This was mainly due to decreases in the populous countries of East and South Asia. Taking China as an example, however, this decline was synonymous with a shift away from cradle-to-grave welfarism in favour of a vast, flexible and precarious labour force, the welfare of which is entirely dependent upon China's economy continuing to create and maintain industrial jobs (Friedman & Lee 2010).

Due to the effective outlawing of migration, another difference regarding the dissolution of the agrarian economy in the global North is that the actively unemployed in the South are territorially contained. There are no external shores to which this growing excess can, at least legally, be exported. Such immobility serves to highlight another singularity of the global precariat: the vulnerability of its off-grid milieus to catastrophe. Between 1973 and 2002, there was a consistent upward trend in natural disasters, as well as their growing economic and infrastructural impact. At the same time, most of the people killed and injured, more than in all other regions combined, have been from 'low-income' countries (UN 2004: 46–8). In 2011, the OECD warned that disasters are likely to become more common and destructive due to the increasing interconnectivity of the global economy and the speed at which people and things now travel (OECD 2011). The openness and vulnerability of the global precariat to hurricanes, floods, earthquakes, famines, health pandemics and conflicts have been repeatedly illustrated in recent decades.

The financial crisis of 2008–9, widely regarded as the worst financial and economic disaster since the 1930s, also demonstrated how the volatility of global markets now impacts directly on the actively unemployed. Capital's easy ability to pass on to a territorially immobile precariat, abject before the uncertainties of its environment, the unmediated force of adverse economic conditions, shortages and market downturns was clear. While the global South is not fully integrated within the financial circuits involved in the

2008 crisis, associated hikes in energy and food costs produced rapid and widespread increases in hardship. Output temporarily fell into negative figures while estimated levels of global unemployment increased by nearly 30 million to reach 205 million in 2009 (OECD 2011: 1). By this time, around a billion people globally were said to be living in hunger, then the highest figure on record. Much of this insecurity, especially increased energy and food costs, has yet to be reduced. The crisis also gave a further boost to the expansion of informality and the spread of economic vulnerability (2011: 4). These estimates lend credence to the consistent upward trend in the number of people globally receiving long-term humanitarian assistance. Defined as for 8 years or more, with medium-term being 3 to 7 years, in 2015 around 88 per cent of official humanitarian assistance went to long- and medium-term recipients combined, with the bulk going to the former. Moreover, of the twenty largest recipient countries, almost all were long- or medium-term recipients (Development Initiatives 2017: 62).

A final consideration helps to broaden this picture. Besides the sedimentation of long-term refugee camps into urban slums, in terms of the growing off-grid world, the urbicidal wars (Coward 2007) being fought in the Middle East and beyond are relevant. Amplifying precedents set in the Balkans at the end of the 1990s, there has been a massive destruction of urban infrastructure in places like Syria, Iraq and Libya. Given the level of infrastructural destruction and dissolution of human capital, one could argue we are witnessing the effective *de-development* of once technologically capable industrial societies. With 11 per cent of its former population holding an advanced degree, Syria was once relatively more educated than the USA (UNHCR 2017b). Having over half of its former population displaced and nearly 6 million refugees beyond its borders, most of its hospitals closed or destroyed, its professional classes scattered and catastrophic levels of privation among the remaining populace, like Gaza, Syria is emblematic of the anticipatory condition of a precariat forced to live amid ruined landscapes. With reconstruction estimates in Syria running into tens of billions of dollars, if the recent past is a guide, we will not see a Marshal Plan-style rescue anytime soon. With the effect of conflict and privation being particularly acute in the Middle East and Africa, according to UNHCR the world is witnessing the highest levels of displacement on record. Of an unprecedented 66 million people forced from their homes, over 22 million are refugees (UNHCR 2017b). There are also 10 million stateless people who, denied nationality and therefore basic rights, have no access to services or, even if they could find it, formal employment.

North goes South

While informalization began first and has spread deepest in the global South (Munck 2013), over the past one or two decades it has also become visible and identified in the North. Indeed, in a reversal of modernist developmental norms, within the third-wave era of electronic globalization, it is the North that is now playing catch-up with the South – catch-up, that is, in terms of the global future of work, at least for the majority, being a precarious and resilience-demanding existence under conditions of permanent economic emergency (Streeck 2011). While there are historical differences, the same post-social logic is reaching and connecting across the mobility barriers that spatially separate the global North and South. In the global North, the casualization of work is associated with the growth of personal debt, at the same time as trade union memberships has shrunk, wages have stagnated, and social mobility declined (Clark & Heath 2015). Some countries, including the USA and UK, are purported to have levels of inequality not seen since the nineteenth century (Roser 2015). It is now on the record that young people have few of the economic life-chances – including predictable employment, housing and pensions – that earlier generations took for granted (Corlett 2017).

That the logic of precarity is now blurring historic global North–South distinctions rests upon a couple of considerations. First, in removing the social protection associated with welfare-Fordism, the new economy brought the North more in line with the more open realities of the global South. Second, the nature of automation in the North is now replicating the 'jobless' growth that has also long been a feature of the South.

Periodic waves of automation are the historic hallmark of capitalist development. While often contested and violent, automation in the past invariably displaced workers into new and dynamic sectors of the economy. Surplus agricultural workers entered the expanding factory system in the nineteenth century. As steam gave way to the combustion engine, and gas to electricity, during much of the twentieth century the diversifying lines of Fordist mass production recycled redundant labour while absorbing increasing numbers of women entering the formal economy for the first time. While the latter had a liberating effect, the necessity of the two-wage, or even multi-wage, family soon became the hallmark of modern precarity. Moreover, as suggested in phrases like 'jobless recovery' or 'jobless growth', concerns have grown in the global North that automation is hollowing out erstwhile professional and middle-class jobs – especially those that depend on logical or algebraic modes of thought. While computers find these jobs easy to replicate, they struggle with low-level sensorimotor skills that rely on mobility and perception (Joshi 2017). Basically, if you want to beat a chess grandmaster, use a computer; to have the pieces cleaned and polished afterwards, find a human. This metaphor gives a clue to the growth of a global precariat.

Automation will not destroy the world of work. It is, however, radically transforming it. The 'good' jobs that require logical and algebraic reasoning – jobs that in the past were associated with stability, career structures, pensions and holiday pay – are declining, being restructured, and concentrating at the top. Most of those made redundant by machine learning, or facing the need to have several jobs to survive, have nowhere to go but an expanding mass of occupations and labour processes that, 'whatever the disparity in wages and skill level among them, have as their common trait that they are *technologically stagnant* ... they exhibit, unlike capital-intensive manufacturing and agriculture, consistently anaemic productivity growth' (original emphasis, Smith 2017). Loosely described as the 'service sector', and having a marked division between business/professional and consumer/personal activities, precarity is subject to varying degrees of casualization and low pay, and is dispersed throughout the retail, leisure and – not least – auxiliary sectors of global manufacturing and agriculture.

The jobs and occupations associated with precarity will not be automated anytime soon. As Jason Smith argues, to believe that this will happen is to assume that the productivity gains recorded in manufacturing due to automation will be easily repeated in the service sector – in other words, 'that a sector resistant to technological innovation and perennially registering minimal growth in labour productivity, will be transformed into dynamic, technologically progressive lines of production' (Smith 2017). To put this another way, it's like thinking that AI and robots will of themselves suddenly transform caring for the growing ranks of the elderly in the global North, chronically unwell, demented and impoverished, into an activity from which capital can extract.

This does not mean, however, that through AI, the precariat cannot be streamlined – that is, as in the 'gig' economy, logistically optimized to be in the right place at the right time 24/7. The global precariat has the appearance of a 21st-century variant of the nineteenth-century servant class (Huws 2015; Boltanski & Esquerre 2016; Smith 2017). While we cannot return to the past, the price of security has been high. Expectations regarding employment protection and what constitutes an acceptable job have been downgraded.

Debates about the changing world of work in the global North commonly focus, often in a celebratory manner, on the growing 'flexibility' of labour markets. For the South, on the other hand, emphasis is on the 'informal' sector, where, as we shall see, a similarly positive revaluation is underway. While flexibility and informality are spatially and historically different, they are treated here as part of a third-wave electronic globalization that is blurring North–South distinctions. Irrespective of its disastrous effects on social reproduction, there is a coming together around the increasing casualization or liquefaction of labour – that is, the creation of bare labour power for which capital has disavowed the social, reproductive and moral responsibilities it had earlier conceded to the external forces of social democracy (Balibar 2016). Helped by advanced business logistics and Big Data, labour's new role is to resemble its early nineteenth-century self on a global stage – to be available and responsive 24/7, easily adaptable and interchangeable, while also being disposable and replaceable without recourse (Reid 2006): as liquid as mobile capital turning a tap on or off at any fixed point on the planet. And, since it is in the global South where the largest and most developed informal labour markets and shadow economies are found (LeBaron & Ayers 2013; Meagher 2016), it is here that the future can be discerned.

Recycling Poverty

Unchecked – if not encouraged – by the NGO-led fantastic invasion of the 1980s, the process of peasant dispossession and impoverishment revealed by structural anthropology during the 1970s (see chapter 4) has continued unabated. By the end of the 2000s, active dispossession and informalization had reached a point that would have been inconceivable three or four decades earlier. Namely, the emergence of a global precariat having little more than its bare, territorially entangled physical and cognitive labour to sell on an uncertain global market. Having given up on social protection decades ago, political and economic elites have obliged the precariat to live on the edge of disaster. In fact, since it is regarded as conducive for resilience, they even recommend it (DFID 2011). Of necessity – and at a terrible human cost, however – the precariat has shown what could be called its actually existing resilience. It has forged and been willing to defend extensive local–global networks and adaptive institutions necessary to support its own social reproduction (Duffield 2001). Despite several decades of 'jobless growth', and with youth unemployment averaging around 40 per cent to 50 per cent, Africa's shadow economies still manage to add some 8 million souls a year to the actively unemployed (Meagher 2016: 485). It is this adaptable, autonomous vitality that a libertarian business sector, wanting to keep overheads down and options high, now seeks to feed on.

Rather than assuming, as in the past, that shadow economies will disappear with development, the sheer size and urban density of the global precariat have forced a rethink. The way that informality is understood, or the 'truth' of informality, has changed. Around the mid 2,000s, views on shadow networking were refracted through the rapidly emerging resilience paradigm (Evans & Reid 2014). By this time, as a means of mitigating disaster risk, the UN was advocating the importance for people in the global South to understand that they are now responsible for their own survival and should 'not simply wait for governments to find and provide solutions' (UN 2004: 189). The precariat, however, had already been doing this for decades. Helped by the resilience perspective, past concerns over informality have been replaced by a new and positive developmental imaginary. Not only are shadow economies inevitable, the fact of surviving and expanding against the odds suggests that the precariat has qualities of innovation and autopoietic self-reproduction that a crowdsourcing, platform capitalism can now, quite literally, capitalize on.

Proposals for what Mike Davis (2006: 79–80, 179) has called 'boot strap capitalism', in which informality has figured as the Third World's potential *deus ex machina,* have been around since the 1970s. With the rise of the resilience regime, however, and the spread of mobile telephony, the idea of a populist capitalism from below has been technologically reinvigorated. Published in 2006, an influential statement reflecting this rejuvenation is C. K. Prahalad's (2006) *The Fortune at the Bottom of the Pyramid: Eradicating Poverty Through Profits*. Prahalad's basic proposition is that we should stop thinking of the poor as a burden, or as victims. Instead, if we 'start recognizing them as resilient and creative entrepreneurs and value-conscious consumers, a whole new world of opportunity will open up' (2006: 1). Bottom of the Pyramid or BOP economics, together with its related ideas of 'inclusive capitalism' or 'inclusive development', highlight the potential value of harnessing this singular low-cost reproductive infrastructure (Meagher 2016).

Orangi Town slum in Karachi, Pakistan, for example, has an estimated population of 2.4 million. Tired of waiting for the authorities to install a proper sanitation system, the residents organized themselves and, by hand, buried sewerage pipes along most of the slum's 8,000 streets. Dharavi in Mumbai, India, houses around a million slum dwellers and, through its tens of thousands of small businesses, provides many services to the city. Its shadow economy has an estimated $1 billion annual turnover. Sited on prime real estate, its residents have up until now resisted official attempts to develop the area (WEF 2016). Rather than fearing informality, shadow milieus can self-organize and reproduce, while providing productive services and low-end mass consumer markets, all within challenging environments. This adaptive vitality now qualifies the informal sector as both a potential development partner and a business opportunity. As Kate Meagher tells us: 'A growing literature on youth unemployment in Africa now argues that what is needed is greater integration of Africa's expanding informal labour markets into local and global value chains in order to increase demand and income opportunities for the vast stores of under-employed labour building up in African economies' (2016: 485).

The aim is no longer to graduate the informal sector to membership of the conventional tax-paying economy. Such formalization would cancel out the shadow economy's post-social resilience when this is the autopoietic quality that is attractive to capital because, for the first time, below-the-radar activities can be revealed and managed through data informatics and remote sensing. In relation to cash transfer, for example, inclusive development envisages using intermediaries like NGOs, social enterprises or labour brokers to go the organizational 'last mile' in integrating informal production, welfare and consumption networks within formal business circuits (Lavinas 2013). In this way, new and hybrid business infrastructures are emerging out of the autopoietic shadow communities created by the precariat to secure their own reproduction. At the same time, this positive orientation towards informality signals a change in the spatial organization of international capitalism. Basically, third-wave globalization is moving out of the enclave and special economic zones into the general milieu of global precarity.

Techno-pastoral 1

Instead of seeing the emergence of a global precariat as an unacceptable affront and evidence of the bankruptcy of capitalism, ideas like 'inclusive development' or BOP economics use a different political compass. The implication is that precarity arose independently, unconnected or – at most – from neglect or lack of attention. It has sprung from the ground as a new object of concern. Now that the shadow economy has been rehabilitated through the lens of resilience, however, rather than the negative talk of 'poverty' or 'victims', the new emphasis is to celebrate the life-affirming vibrancy purportedly revealed – especially a vitality, inventiveness or resilience that is drawn out and enhanced by new technology. We have called this the techno-pastoral aesthetic.

This aesthetic is a contemporary developmental inflection of the 'progressive neoliberalism' (Fraser 2017) outlined in chapter 2. Namely, the fusion of demands for market freedom, rights, choice and authenticity with a Hayekian neoliberalism that, rather than around *Homo economicus*, is fashioned towards the distracted and cognitively challenged *Homo inscius* (Spencer 2016a). Chapter 5 broadened the idea of progressive neoliberalism in relation to the NGO-led fantastic invasion of the 1980s. Developmentalism resides in the various relations of care, inclusion and giving voice that interconnect progressives with the *Homo inscius* of contemporary development. Regarding the fantastic invasion, through the project form, progressive neoliberalism was instrumental in the valorization of community mutuality and the relations of social reproduction for the marketplace, so to speak. Today, it is playing a similar anticipatory role, this time regarding the incorporation of precarity. As argued in the next chapter, connectivity has enabled a moving beyond the project form towards the incorporation of precarious life generally within the circuits of global capitalism.

Progressive neoliberalism has aesthetically appropriated the shadow networks of precarity in the global South in the form of a romanticized poverty bucolic. While this aesthetic is examined further in the next chapter, here it is sufficient to note how the informal sector is idealized as a living embodiment of indigenous knowledge, local management practices and active community mutuality, including making space for and valuing women (Becker 2004). Moreover, the very act of avoiding conventional rules and regulations qualifies shadow networking as a surrogate resistance to 'neoliberalism' itself (Jackson 2016). The techno-pastoral aesthetic invests precarity with vibrancy, authenticity and hope. Through this progressive reinscription, informality reappears as an eager and eligible development and business partner. Consider, for example, UNDP's homely appraisal of NGO-assisted informality as a low-cost welfare infrastructure for an inclusive capitalism:

> A community is more than the sum of its parts. Where poverty prevails, formal laws and regulations are often less effective than the informal rules that communities set and enforce. Such informal rules can make inclusive business models viable. And a community can help its members to help each other – for example, by sharing resources, by co-operating to provide common goods (such as wells, mills or schools) and by supplying an infrastructure for savings, credit or insurance mechanisms. Businesses can count on these communal processes to fill gaps in the markets of the poor. (UNDP 2008: 9)

Apart from an undercurrent of paternalism, there is a peculiar bipolarity that haunts such bucolic depictions of poverty. Given that precarity is now a global phenomenon, any celebration of informality in the global South should, by implication, also resonate with casualization in the North. The techno-pastoral aesthetic, however, doesn't translate. At the time of writing it is difficult to imagine that UNDP could publicly lobby international business regarding, for example, the free-rider attractiveness of the UK's expanding gig economy, pointing out, perhaps, its flexibility, lack of unionization, acceptance of zero-hour contracts and, importantly, that community-run food banks can be counted on to 'fill gaps' when remuneration or the benefit system fail. This 'lost in translation' quality is symptomatic of an enduring spatial psychosis that appears to be intrinsic to developmentalism. A lordly bipolarity presides over a situation where what's inappropriate or difficult in the global North, like the anticipatory structural adjustment of the 1980s, is necessary or doable in the South.

Linnet Taylor, in describing the 2013 Internet Governance Forum, gives an example of this psychosis in action among its participants. Taking place in the wake of the Snowden leaks, concerns about data protection and internet privacy in Europe and the USA were important talking points and high on the conference agenda. However, whereas, '90% of the discussion at the forum referred to big data as a tool for surveillance ... the thread of debate that focused on developing countries alone, *treated it as a way to "observe" the poor in order to remedy poverty*' (emphasis added, Taylor 2013; also see Pirlot 2014).

Developmentalism combines a paternalistic, techno-pastoral aesthetic with the relative licence that the global South affords. This should not be confused with 'Eurocentrism' (Sabaratnam 2013). The progressive neoliberalism underlying the techno-pastoral is adamantly cosmopolitan: this is its strength. As during the colonial period, the post-colony remains a site of the boomerang effect. Its lack of regulations, pervasive state and business corruption and, importantly, its weak data and privacy laws (Hosein & Nyst 2013), besides being the conditions of informality itself, mean that the global South remains a site of commercial and humanitarian anticipation and experimentation (Jacobsen 2015). There are several factors, however, that are specific to the moment. The rapid penetration of mobile telephony and its associated data informatics promises to do what conventional economic regulatory tools and policing methods never could – that is, to expose and bring the flows, exchanges and networks of a now-surveilled precariat into the light of day. Moreover, this uncovering and exposure is not to curb informality, or graduate it to the formal sector. These aims have now disappeared. The current intention is to capture precarity, and to encourage and exploit its disruptive beyond-the-law potential.

The next chapter examines how smart technology levels downwards into the social fabric, enabling capitalism to move out of the special economic zone to embrace and incorporate precarity as a whole.

Chapter 9

CONNECTING PRECARITY

As a mode of life beyond the social – that is, beyond the former realities of modernity in the global North and its promise in the South – precarity is a historically novel form of dispensation. It reflects an expanding life-world that exists at the interface between economy and disaster. Post-humanitarianism encompasses the data-based interventions, design frameworks and ethical orientations geared to augmenting precarity in contexts where political change has been negated by resilience, and where a progressive neoliberalism is the dominant window on the world. It embodies an anticipatory, socially disaggregated biopolitics that, through datafication and remote sensing, seeks to optimize logistically the body's fitness, recuperative and cognitive capacities sufficiently for smart humans to exist in the post-social wild. Long anticipated in artistic and critical thought (Gibson 1984; Hayles 1999; Dillon & Reid 2009), post-humanitarianism is the realization within the laboratory of the global South of a biohuman essence that blurs the distinction between mind and market, and human and nonhuman systems.

Third-wave globalization, to which post-humanitarianism belongs, is deepening the break with modernism that the fantastic invasion of the 1980s began. Post-humanitarianism denotes the automation of aid, together with the embrace of remote management and the new commercial and private actors that are inseparable from these technologies. Rather than automation being exceptional, humanitarianism is following a consistent trend for an ever- widening range of previously grounded knowledge-based activities to be captured by what Bernard Stiegler has called *automatic society* (Stiegler 2016). Apart from the computational turn and rapid diffusion of mobile telephony, humanitarian automation builds on several key developments. Besides austerity pressures to reduce costs, an important justification has been the risk-related retreat of international aid workers, journalists and academic researchers from challenging environments and disaster zones. However, as a hollowed-out and restricted engagement with the outside world has become naturalized, a much more positive and rejuvenating ontological force has been harnessed. Following their mauling by the May '68 movement, there has been a rehabilitation and subsequent rise to dominance within the academy of positivist, empiricist and, not least, behaviourist strands of thought, under the rubric of *post-humanism* (Braidotti 2013).

While approaches like speculative realism (Harman 2010), the new materialism (Coole 2013) or actor network theory (Latour 1987) have internal differences, reflecting Marcuse's (1968) critique of technological society, they tend to privilege flat, process-oriented ontologies of becoming that cast individuals as relationally embedded within the pure factuality of their immediate environments (see Galloway 2013; Chandler 2015). Rather than a dualist separation between individual or concrete 'reality' and a higher-order 'world' – accessible through theory and open to critique – life is more an exercise in pure or unmediated factuality, as monadic individuals constantly connect and disconnect from others and things across horizontal data landscapes. As David Chandler has argued, by bringing the interactions and interrelationships between the semiotic and the material to the surface, thus making them 'readable and thereby governable', Big Data has allowed post-humanism to come of age (Chandler 2015: 838). Paradigmatically, this ontology achieves a clear expression with regard to the ordering effects of the screen interface. No longer dependent upon circulation, the connected 'world' becomes so many fragmented and personalized linear connections between friends, feeds and member forums. As Chandler points out, post-humanism's coming of age can be measured in its effortless passage from critical outlier into the policy mainstream (2015: 849). Big Data plays an enabling role in the sense of authoring post-humanist 'ways of governing the world based upon process-based understanding and relational ontologies' (2015: 838).

Post-humanism gives intellectual coherence to post-humanitarianism. In beginning to explore this coherence, this chapter examines from an infrastructural perspective the expansion of mobile connectivity among the global precariat. If connectivity has enabled the flat ontologies, pure factuality and behaviourism of post-humanism to come of age, it has also underpinned a major shift in the spatial organization of network capitalism. From an earlier reliance on fenced-off special economic zones that formally exempted capital from the law, third-wave globalization and the datafication of the vast informal economies of the global South constitute a movement beyond the enclave, rendering society as a whole an economic zone of exemption and disruption. At the same time, however, in terms of the techno-pastoral aesthetic, there is a curious feeling of smart technology presiding over unchanging landscapes of poverty and decay. This is the terrain of post-humanitarianism.

Levelling Downwards

Taken together, the existing physical world of wires, buried pipes, dams, power stations, telecommunications networks, transport links and urban architecture constitutes an engineered environment that traditionally functioned, often in the background, to maintain circulation. The foundations for much of today's critical infrastructure and urban design were dug during the period of capitalist acceleration prior to the 1980s (Graham & Marvin 2001). Before its extensive privatization, this infrastructure could be seen as a large-scale, fixed-capital technical grid (Balakrishnan 2009). At the height of the welfare-Fordist period of mass manufacture, this universal fixed-grid, together with its inbuilt redundancies, standardized tariff systems and universal connection protocols, was associated with nationalized or state-regulated essential services such as transport, water, energy, public housing, health and telecommunications. The fixed-grid aimed to provide public access to a universal set of standardized utilities and services through fixed tariff regimes. It marked a time when rich and poor, so to speak, were connected to the same water, electricity and sewage systems. Its growth since the nineteenth century is inseparable from a political economy of urban modernism and the improvement of living standards through normative interventions concerning health, education and employment conditions (Rabinow 1995). Prior to the 1980s, in the erstwhile Third World, attaining a similar universal fixed-grid of utilities and services embodied the modernist aspirations of states when development still meant economic catch-up with the West (Rostow 1960). In aspirational terms, the universal fixed-grid promised a collective *levelling-up* for society as a whole.

The infrastructure supporting a fast-expanding global connectivity includes the satellites, fibre optic cables, towers, routers and wires that interconnect a growing network of climate-controlled data warehouses with billions of roaming screen interfaces. This infrastructure also constitutes a political economy (Terranova 2004; Lesczynski 2012; Easterling 2014).[1] Rather than directly replacing the old universal fixed-grid, this data-hungry global ecosystem acts upon, reinvents and transforms it. Rather than renewal as such, harnessed to neoliberalism the spread of connectivity has been associated with the privatization of the fixed-grid, the undoing of social or normative welfare, and the commercialization and marketization of the life-worlds it supported (Spreeuwenberg & Poell 2012). As reflected in the phenomena of jobless growth and increasing inequality, privatization has largely been parasitic on the universal fixed-grid – fragmenting and globalizing while under-investing in repair and replacement as it feeds zombie-like on the dead labour it has acquired (Balakrishnan 2009). Of itself, increased connectivity has not closed the growing gap between global requirements and the slow pace of infrastructure renewal and reconstruction (Mckinsey Global Institute 2016). In terms of the sprawling slums and ruined landscapes in question, this gap directly impacts the global precariat.

Technoscience and the business world have responded to the infrastructural gap through the ethos and medium of smart technology. Rather than renewal *per se*, smart technology is more a replacement that operates through bricolage and the leverage of existing infrastructure through the development of new business models and marketing strategies. Such leverage has reached Byzantine proportions, for example, in relation to the UK's railways, domestic energy and broadband supply. A privatized fixed-grid has provided a foundation for the creation and regulation of several artificial markets. While the state may have shrunk, an expanding network of private providers, confusion marketing and diminishing public accountability have more than compensated (Agamben 2013). At the same time, the administrative and transaction costs that were once absorbed by business and state sectors have now been passed to customers.

In urban development terms, smart technology has facilitated a move away from normative city planning towards selective gentrification, gated communities and the privatization of public space that draws physical and cognitive lines between better-off areas and the food and amenity deserts of the precariat (Davis & Monk 2007; Minton 2009; Spencer 2016a). Contrary to the levelling-up logic of a universal fixed-grid, smart technology levels *downwards*; it folds into and reproduces the varieties, differences and inequalities within the human terrain. In relation to these differences, connectivity provides a sense of democratization and empowerment. Indeed, it proclaims as a universal right that smart technology should engineer different speeds, tariff bands, access protocols, technical fixes and customized packages according to the societal inequalities and varying bandwidths encountered. As Peter Redfield points out, smart technologies in the global South 'must adapt to an absence of support infrastructure. They must survive a perilous environment and cannot depend on a regular supply of electricity. To be successful, humanitarian goods must recognize their users, adjusting to the reality of their worlds even as they seek to change them' (Redfield 2015: 15).

Smart technology folds downwards into the social fabric. It codes, maps and digitally reproduces the startlingly unequal fitness landscapes of network capitalism. Like resilience, however, smart technology is not designed to eradicate the root problems that lie behind the inequalities encountered. On sliding scales of cost and effectiveness, it promises 'the connected' ways of sidestepping ground friction as they navigate the old circulatory spaces of a residual urban modernism that is now the new wild.

Electronic atmosphere

Integrating the vast shadow systems of precarity within global value chains represents a complex challenge for business logistics. That such a challenge is conceivable, however, rests upon the rapid diffusion of mobile information technology across the global South. Since the mid-2000s, through subsidies, engineering innovation and software for data deals, cheap mobile telephony has, to use a phrase from the early days of satellite coverage, 'leapfrogged' over decayed or absent terrestrial telecommunication systems (Skinner 2010). According to the World Bank, mobile telephony is now the largest distribution platform in the world. Between 2000 and 2012, the global number of devices rapidly expanded from 740 million to nearly 7 billion, with three-quarters of these located in the global South (Easterling 2014:

17). Today, even in remote areas, the precariat have access to at least some bandwidth through basic mobile devices (de Bruijin et al. 2009; Donovan 2013; Nielsen 2015). This deployment is of world-historic importance. To get a sense of its effects and potentialities, it is worth considering leapfrogging not as a horizontal process of technology transfer but as a prime example of smart technology levelling downwards. As Kelly Easterling has pointed out, while the prospect of Development 2.0 is widely celebrated, 'the discipline is under-rehearsed in an analysis of the spatial dispositions attending broadband infrastructure' (Easterling 2014: 97).

During the decade following the late 1980s, most of the world was connected by fast, undersea fibre optic cable. In terms of its political economy, this surge of digital connectivity stands comparison with the laying of the analogue submarine telegraph cable network during the nineteenth century (Headrick 2012). Fibre optic cable came relatively late to Africa. From less than ten landfalls by 2009, mainly along Africa's west coast, by 2012 this grew rapidly to more than thirty, which now encircle the continent (Song 2015). Before connection, East Africa, for example, represented only 1 per cent of global broadband capacity (Easterling 2014: 95). The region had earlier relied on older and expensive satellite technology, for which the auxiliary infrastructure was created in the 1970s. Between 2009 and 2010, while of relatively low capacity, three commercially owned cables made landfall at Mombasa on the Kenyan coast (2014: 113). As with the rest of the continent, these cables are buried in the ground and generally follow the main arterial routes from the coast to the interior. Looking at a broadband trunk cable map of Africa,[2] one is struck by its similarity to the topology of the colonial railway system. With the exception that fibre optic cables now often cross frontiers, they are similar in usually being single-tracked, moving inward from the main ports, and linking a few principal towns and main export centres while carving out huge intervening white spaces. The difference from an average European or American city is striking. Here, fibre optic cables are locally and densely interwoven, buried along railway lines, main roads, streets and cul-de-sacs in a bid to connect as many individual homes, schools, offices and businesses as possible. In addition to the primacy of battery-powered devices, this infrastructural difference suggests that last-mile broadband connectivity across Africa's huge electronic white spaces remains, and will continue to remain, an atmospheric problem.

In Kenya, the main trunk cables are buried alongside the Mombasa–Nairobi highway. While two-lane for much of its length, its surface is poorly maintained and difficult to navigate in places. This reflects the paradox of the diffusion of mobile telephony: it exists alongside high levels of precarity and the neglect or disrepair of auxiliary infrastructure (2014: 98). While a few enclaves are connected along the way, in the main, the cables make for the capital and beyond. Relying on these trunk cables for bandwidth, mobile telephony adds a new 'atomized typology of microwave towers and handsets' (2014: 98). Interconnecting battery-powered devices, these numerous towers create an electronic atmosphere of varying density stretching out from points of access to the high-speed fibre optic backbone. Compared to earlier satellite technology, this surface atmosphere of microwave towers greatly increases last-mile connectivity and roaming ranges through the medium of battery-powered mobile phones (2014: 97). The mainly local or regional service providers that sell and transfer broadband between the overlapping systems of cables and towers add another infrastructural layer comprising 'a cluster of switches or points of access' (2014: 97). Monopolies, bottlenecks and competition can develop 'within these linear, atomized, and clustered topologies' (2014: 97).

Besides local and regional telecommunications companies, Silicon Valley is also involved. This includes the retrofitting, or levelling-down, of corporate software platforms to work in the low-bandwidth conditions encountered. Retrofitting involves lower-density graphics, advanced caching techniques and improved batteries to optimize off-grid roaming (Honan 2014). Moreover, in their bid to capture the 4 to 5 billion people globally still unconnected – a billion of whom are said to be in Africa – Google and Facebook are both investing in new stratospheric communication infrastructures operating at lower altitudes than satellites (Naughton 2014). In the words of Mark Zuckerberg, this willingness to adjust to poor tele-economic conditions is an indication of Facebook's desire to develop new platforms 'based on the principle that different communities need different technical solutions' (Zuckerberg 2014).

In order to reduce costs, and to avoid terrestrial regulations and, not least, the potential insecurity of ground installations, interconnecting satellites, trunk broadband cables, relay towers and millions of roaming screen interfaces from the stratosphere has advantages. Google, for example, is exploring high-endurance balloons capable of drifting well above the Earth's surface, using stratospheric winds to navigate (Google 2015). Facebook is investing in high-altitude solar-powered drone technology, with the aim of creating unmanned vehicles able to stay aloft for months at a time, rather than weeks or days (Zuckerberg 2014). Called Aquila, after the Latin for eagle, the first successful test flight took place in May 2017 at the Yuma Proving Ground. With the wingspan of a Boeing 747, fleets of these drones will eventually, according to Zuckerberg, 'beam internet connectivity across the world' (Associated Press 2017). These rival technologies are similar in that they seek to operate from the stratosphere, or more than 20 kilometres above the Earth's surface. Not only is the stratosphere well clear of ground friction, from a legal perspective it is like the high seas. It constitutes an ambiguous space that is, literally, above the law (Neocleous 2013). Unlike the ground versions of 'ungoverned space', however, which are prone to insecurity, political push-back or access denial, the stratosphere is being colonized by the digital corporations. Even the laser-beam technology (free space optics communications, or FSO) that is being optimized for fibre-like speeds of data transmission between drones and the terrestrial tower network uses a part of the radio spectrum that is outside international telecommunication regulations (CableFree 2017).

Expanding the Enclave

Claims for the leapfrogging potential of atmospheric connectivity in the global South are not new. In the 1970s and 1980s, futurologists imagined that satellites could bypass the terrestrial fixed-grid by simply beaming down a communication infrastructure (Easterling 2005: 136). In practice, however, satellites could not solve the last-mile problem. The necessary auxiliary equipment required a fixed electricity supply and dedicated fibre and cable access points. For such reasons, satellite technology favoured the creation of well-resourced infrastructural enclaves that serviced urban elites and took the form of office parks and special economic zones (2005: 138). Mobile connectivity radically changes this socio-spatial configuration. Bottom-of-the-pyramid economics signals a shift in the nature of global capitalism. Third-wave globalization involves moving beyond the off-shore manufacturing enclaves and special economic zones that underpinned Northern deindustrialization during the

1980s and 1990s (Amsden 1990; *Beijing Review* 1992). The special economic zone was an authoritarian economic construct typical of second-wave globalization. Such enclaves are usually 'an isomorphic exurban enclave that, exempt from law, can easily banish the circumstances and protections common in richer forms of urbanity' (Easterling 2014: 15–16). Special economic zones are fenced-off and securitized spaces where national tax liability, safety requirements or investment regulations either do not apply or can be negotiated in the interests of business (Cowen 2014). The zone is where bare labour enters to meet the external investment it has attracted.

In the wake of the 2008 financial crisis, based upon the infrastructural platform of mobile connectivity, capital is now reaching out beyond the enclave to embrace the precariat. In so doing, it is discovering in the informal sector another environment that, while different in some respects – especially in terms of its vast low-end consumer potential – is also structurally similar to the special economic zone. Shadow economies are *de facto* also 'exempt from the law'. As with the behaviour of digital corporations, avoiding taxation and national regulations has long been a defining feature of informality. Through incorporating the social reproduction of the precariat, third-wave globalization from the electronic atmosphere has the potential to transform the global South as a whole into a vast 'special economic zone'. Given that shadow networks are valued for their autopoietic powers of production, reproduction and consumption, and that these are possible in part because of their self-exemption from tax and the law, one can expect electronic globalization to deepen the casualization of work (Meagher 2015; Meagher 2016). One could go further, and join Genevieve LeBaron and Alison Ayers (LeBaron & Ayers 2013) in recognizing in casualization the logic of a modern slavery – modern in the sense of breaking the traditional ownership link between master/slave, yet still riven by abjection. A post-social capitalism breaks all ties of responsibility between finance capital and labour, while connectivity helps to manage the huge distances and multiple organizational relay points that now separate them. If we can use the term 'modern slavery' in relation to the precariat, it refers to the abject dependence that is generated by a capitalism that now has the power to hold labour in a state of permanent emergency.[3] Repacked as development, with all the celebratory hype this entails, capital's absorption of a self-reproducing precarity looks set to expand.

In Kenya, which is not untypical, the explosion in mobile phone usage is remarkable. Pricing is so low that, even before the landfall of broadband, devices were socially widespread. Between 2000 and 2008, subscriptions increased over a hundredfold, from 127,000 to around 16 million. With the arrival of broadband, they nearly doubled again between 2008 and 2012, to some 30 million devices among a total population of 48 million (Easterling 2014: 96). However, to draw out the real significance of this increase in connectivity, it needs to be flagged that Kenya has one of the largest informal sectors in Africa (UNECA 2015a: 66). Involving around 12 million people, informal employment in Kenya represents 83 per cent of the total. This includes small retailers, street vendors, unlicensed motorcycle taxis and casualized employment in the agricultural and manufacturing sectors. Women are particularly involved in shadow trading networks. Formal employment, providing a regular wage or salary, has continued its consistent decline, resting at just over 4 per cent of the workforce in 2014 (2015a: 16). Varying widely according to location, in 2009 the average level of extreme poverty was estimated at 45 per cent (2015a: 24). According to the UN's Economic Commission for Africa, having no formal safety nets, the Kenyan precariat 'operate under a high degree of informality and vulnerability, resulting in small and unpredictable income, poor working conditions and low productivity. Such informality is likely to trap people into poverty' (2015a: 67).

Nairobi has 2.5 million slum dwellers, comprising some 60 per cent of the total urban population, yet they live on only 6 per cent of the land. Kibera is the largest of Nairobi's informal settlements and is the largest slum in Africa (Kibera 2017). Overcrowded, with 90 per cent of the population having no tenancy rights, there are no government clinics or hospitals. Having a wide range of health issues and lack of education provision, basic services are provided by NGOs. Only 20 per cent of the settlement has electricity, and potable water for sale is supplied by access points off two piped supplies. Other than shared pit latrines – in some cases, dozens of overcrowded shacks to one latrine – there is no fixed sewage system. Unemployment is high despite a widespread involvement of slum dwellers in the informal sector (Kibera 2017).

Not all Kenya's precariat live in slums or earn less than $1.90 a day, the current UN marker of extreme poverty. Moreover, there are gaps in the statistical data and a danger that such an overview ends up comparing apples with oranges. However, given that Kenya is not an exception with regard to the intermeshing of informality and connectivity in the second decade of the twenty-first century, these estimates give some indication of the historically singular milieus that are emerging in the global South. Vast and populous urban and peri-urban post-social zones, themselves vulnerable to disaster and external shocks, where a connected and actively unemployed autopoietic precariat is effectively contained. Having few if any assets, its coping strategies, and physical and cognitive labour, are realized within informal and insecure global markets increasingly mediated and controlled by digital logistical platforms. However, using the moral compass of inclusive capitalism, this development is not viewed as a failure or affront. Just as the dominance of atmospheric connectivity establishes an infrastructural difference between North and South, third-wave globalization is creating a new set of anticipatory possibilities among the milieus that capitalism's crisis of social reproduction is creating.

In 2010, the World Bank was already arguing that the model of market liberalization that had hitherto been followed in Europe and the USA 'is not directly relevant to the region of Sub-Saharan Africa' (quoted by Easterling 2014: 108). As in Kenya, a new strain of liberalism is emerging:

> a liberalism associated with the platforms of exchange made possible by new technologies. In countries like Kenya the low prices and large customer volumes of mobile telephony align with the new 'trickle up' business and management models emerging from populous countries of the global south. The idea is to sell a limited inexpensive service or product like the cell phone to a large number of people. (2014: 108)

Kenya is being hyped as a potential 'silicon savannah', and broadband is now written into government policy and development goals (2014: 97). Entrepreneurs are trying to identify multipliers, find new bottom-of-the-pyramid business models, refine crowd-sourcing techniques and develop apps to capture the producer and consumer markets that the precariat, due to its increasing size and low overhead costs, represents. As Easterling notes, companies like TATA, MTN, Safaricom and Huawei have so far been bypassing their Western counterparts in developing such approaches. However, reflecting the relatively low capacity of Kenya's undersea fibre optic cables, while atmospheric connectivity is adequate for 'boot strap' capitalism, despite the Development 2.0 hype, the country is 'sorely lacking high-capacity premium fixed bandwidth for business and education' (2014: 96). In other words, bottom-of-the-pyramid economics is just that. It continues the trend established in the 1970s and 1980s of denying, for one reason or another, that the global South is suitable for what used to be called a modern industrial future, with all its implications.

Current moves can be condemned as threatening to digitally reproduce anew the old development–underdevelopment divide. However, there is a danger of missing the point. Experiments in recycling and streamlining the precarity and insecurity that the new economy has produced in the global South is where the future of capitalism, North and South, is to be found. While critical attention is rightly directed towards the consequences of deepening social automation in Europe and the USA (Stiegler 2016), the future of network capitalism will be equally defined by its ability to valorize the global precariat in all its 'bandwidths'. Whether the enterprise will succumb to the anger, violence and resistance that austerity first engendered (Walton & Seddon 1994; Bush 2007), and which increasing casualization is deepening (Mishra 2017a), is an open question. It does not instil confidence, however, to realize that Silicon Valley elites are preparing for the future by buying properties in New Zealand (Donnell 2017) or constructing and furnishing wilderness hideaways (Bartlett 2017).

Techno-pastoral 2

The tension between the people and things that have leave to circulate, and those that are territorially immobile and denied movement, is now a major security issue along the globally expanding, heavily policed and electronically surveilled migratory interface of barriers and checkpoints (Brown 2010). The spatial and temporal asymmetry between mobility and immobility creates a number of post-humanitarian demands for otherwise restricted or hyper-bunkered aid managers – for example, the ability to connect with distant and contained populations, for humanitarian commodities or technologies that can be self-administered and operate off-grid, and for means of remote evaluation and identity verification. There is a demand for cheap humanitarian technologies that are conducive to both aid automation and remote management. The issue of humanitarian innovation is examined more fully in chapter 11. Here, it is sufficient to say that its logic follows that of smart technology. Humanitarian design adapts to absences, inequalities and speed differentials by folding into them, while, at the same time, reproducing them. Preferring design over revolution, the dominant aesthetic of post-humanitarianism is the already-mentioned techno-pastoral (see chapter 8).

An early example of the techno-pastoral aesthetic can be traced in Alvin Tofler's *Future Shock* published in 1980 (Tofler 1980). In a chapter aptly entitled 'Ghandi and Satellites', this prescient work foresees technoscience as providing a third wave of development that resolves the contradictions between a first-wave agrarian traditionalism and second-wave industrialization. Satellite and other advance technologies, rather like today's mobile telephony, were seen leapfrogging over existing first-wave conditions and impediments, enabling betterment without full industrialization and thus the need to sacrifice cultural values, communal mutuality, religion or indigenous practices: 'Given the wider range of options brought by the Third Wave, cannot a people reduce infant mortality and improve life span, literacy, nutrition, and the general quality of life without surrendering its religion or values and necessarily embracing the Western materialism that accompanies the spread of Second Wave civilization?' (1980: 337). Advanced technoscience would enable local forms of development that would not overemphasize the economic factor 'at the expense of ecology, culture, religion, or family structure and the psychological dimensions of existence' (1980: 337; also Easterling 2005: 136–8). While Tofler was not able to foresee the widening inequality that progressive neoliberalism would produce, it is interesting that, nearly four decades on, humanitarian innovation remains trapped within what is now an old futurology (Barbrook 2013).

Rather than radical change – indeed, this is rejected – humanitarian innovation celebrates how technoscience has fabricated a range of digital infrastructures, humanitarian objects and smart technologies that enable the precariat to survive off-grid within essentially unchanging poverty landscapes. Countless photographic images, installations and promotional brochures attest to the difficulties that post-humanitarianism has in visualizing progress. Instead, we find a recurrent techno-pastoral aesthetic where timeless scenes of poverty, slum dwelling and everyday survival are product-placed with mobile phones, solar-powered lamps, flat-pack shacks or other examples of humanitarian design. Complete with smiling people, these are so many homely tableaus of Halpern's (2017) 'hopeful resilience'. Since Tofler's day, the only difference is that the slums are now wired. The techno-pastoral aesthetic is forcefully expressed in the work of the photographer Ruben Salgado Escudero, whose series of stylized 'solar portraits' of the precariat 'within their environment' condense a discomposing feeling of progressive entropy.[4] Whereas Tofler saw the third wave as offering a dignity to traditional values, and community mutuality that remained independent and autonomous, a parasitic inclusive capitalism, while it praises the same autopoietic qualities, is now incorporating and absorbing them.

The remainder of this book examines the essentially entropic logic of post-humanitarianism. The next chapter begins this task with an analysis of remote satellite sensing and the emergence of data informatics as the new tools that now make sense of precarity from a distance.

Chapter 10

POST-HUMANITARIANISM

The infrastructural spread of global connectivity roughly breaks down into two interconnected phases. The first concerns the expansion of commercial satellite remote sensing from around the mid-1990s, allowing innovations in humanitarian remote sensing to take place prior to the rapid spread of mobile telephony. This included developing algorithms for timely data extraction from high-resolution imagery of refugee and other disaster-affected groups. Important here were innovations in the visualization of such data, in the form of layered geographic information system (GIS) maps. The second, or current phase, sometimes known as Web 2.0 (O'Reilly 2005), gathered momentum from the mid-2000s with the rapid spread of broadband connectivity and geolocated mobile telephony. This has significantly enhanced the governmental potentialities of digital technology. Rather than simply being sensed from afar, the precariat have become active producers of exponentially expanding volumes of data in their own right. Since the latter part of the 2000s, emergencies have been real-world laboratories for innovations in data-based sense-making (Meier 2015). The ability to combine remote sensing from space – and, increasingly, from ground sensors (Gabrys 2016) – with mobile data informatics constitutes post-humanitarianism's conditions of existence.

Digital connectivity is both infrastructural and ontological. As well as being an engineered system of satellites, towers, cables, data warehouses and screen interfaces, this mnemotechnic, or recording and memorizing infrastructure, calls forth post-humanist, behaviourist and cognitive ontologies and methodologies. These technologies encourage remoteness from the world, in as much as they spare practioners, journalists and researchers from having to take risks. At the same time, in complex and fluid environments, they ease the burden of thinking and logistical decision-making. They make the automation of humanitarian assistance in challenging environments possible. With an emphasis on the digital recoupment of distance and the central role played by *Homo inscius*, this chapter outlines the development of remote sensing and data informatics in the humanitarian field.

Remote Sensing

The increasing use of remote sensing by the aid industry since the end of the 1990s should not be seen as a wholly new departure. Certainly, the actual physical technologies had not been used before. However, in terms of an expanding cybernetic *episteme*, chapter 5 explored how the NGO-led fantastic invasion of the 1980s was instrumental in the anticipatory transformation of social knowledge into behavioural data. Emergency events were reinterpreted from an 'early warning' perspective as emergent complexes of measurable signals and behavioural alerts. Rather than area expertise or tools of historical or social analysis, these signals demanded new field methodologies to record, analyse and initiate timely action. Geospatial remote sensing built on this earlier anticipatory analogue foundation. The view from space gave the necessary distance for greater objectivity.

Like the internet (Lukasik 2011), the application of remote sensing to complex emergences is intrinsically linked to the civilianization of military research and development. It is easy to forget that, as recently as the mid-1990s, high-resolution satellite imagery, the associated sensor technologies and GPS positioning were still classified and restricted (Verjee 2005).[1] In the mid-1990s, as part of the wider deregulation, privatization and globalization of the universal fixed infrastructural grid (Collier & Lakoff 2007), the Clinton administration introduced several crucial geospatial[2] reforms that allowed the rapid privatization and commercialization of military technologies and data-sets (Verjee 2005). Beginning in 1993, initially a military creation, the satellite-based Global Positioning System (GPS) was progressively opened to civilian use.[3] Today, mobile phones can potentially be logged to within several metres at any point on the Earth's surface. The declassification of military imaging sensors was followed by the authorization of the commercial operation of high-resolution satellites (Hayes 2012). Although the military retains a technological edge and, if necessary, 'shutter control' through funding and licensing agreements, civilianization has in many respects reversed the dependencies of the Cold War military-industrial-academic complex in favour of the private and corporate sector.

The military now regularly enters into commercial contracts to purchase geospatial products from independent suppliers, as in the case of commercial satellite imagery of Afghanistan at the start of the international intervention (Ackerman 2001). Civilianization has also been coterminous with a change in military sensing priorities from those that characterized the Cold War. Rather than scanning the Earth for objects and buildings, such as missile sites or armament factories, the main concern now is hunting people (Chamayou 2015 [2013]). This has accompanied a shift towards large-aperture/ low-power devices, together with an increased use of computer-generated 3D visualizations (Hayes 2012). This recalibration around the extraction of human features and behavioural data means that the military, humanitarian agencies and human rights organizations are now fellow customers within the same technology marketplace.

During much of the 1990s, humanitarian organizations struggled with the costs of remote imaging and accessing the expertise to interpret it. By the end of the decade, however, led by the UN, especially UNHCR and ICRC, the aid industry was regularly using and experimenting with geospatial technologies. These were accessed through partnerships with public–private space consortia like UNOSAT[4] and RESPOND,[5] or state-supported research networks like the EU's Joint Research Centre, or JRC (Hanchard 2012).[6] Such first-wave consortia typically brought together, in varying combinations and different terms of agreement, national space agencies, university research facilities, private geospatial companies and the UN or other humanitarian agencies (Verjee 2007). These geospatial partnerships aspired to leverage private-sector technology, expertise and data philanthropy to meet urgent humanitarian needs. This was the background to UNHCR's first exploration of the use of geospatial technology for refugee management (Bouchardy 1995). By 2000, a detailed geospatial mapping exercise covering refugee camps in Kosovo, Nepal and Kenya had been completed in partnership with

ENVIREF,[7] a consortium of European geospatial companies (EnviRef 2001). This has been claimed to be the first systematic application of geospatial technology to complex emergencies (Verjee 2005). It successfully demonstrated the feasibility of extracting humanitarian intelligence from commercial high-resolution images of refugee camps and their environs.

Using this experience, at the onset of the 2004 Darfur emergency in Sudan, UNHCR asked UNOSAT to search for groundwater reserves in Chad sufficient to support Darfur refugees (Bally et al. 2005). Within months, the consortium had procured accurate water supply target maps, which UNHCR used to optimize the location of some of its Chad camps. As the emergency deepened, by December 2005 there were 1.5 million Internally Displaced Persons (IDPs) within Darfur. UNHCR commissioned RESPOND to supply thematic mapping and route-planning data. Over the next couple of years, besides humanitarian activity maps for the United Nations Office for the Coordination of Humanitarian Affairs (UNOCHA), RESPOND supplied a range of cartographic and logistical products to donors, UN agencies and NGOs (Donnelly 2012). These included area resource maps, aerial camp photography, water availability data, transportation routes, etc. Using ten different sensors from nine spacecraft, RESPOND extracted geospatial data under a number of public–private agreements (Hanchard 2012).

Since the mid-1990s, refugee camps have been increasingly militarized and securitized, and local restrictions on international access have grown. In Darfur, for example, many camps are controlled or contested by opposition groups (Kahn 2008). Refugee and IDP remote sensing has grown in scope and accuracy in concert with growing restrictions on humanitarian access (HERR 2011). The scientific literature on the geospatial dimensions of the Darfur emergency frequently cites the danger on the ground for international aid workers (Sulik & Edwards 2010). From the start of the emergency, the EU's JRC worked on the problem of remotely measuring Darfur's volatile IDP populations. By the end of the 2000s, scientists had developed algorithms to do this automatically and to acceptable accuracy, without the need for ground truth (Kranz et al. 2010; Lang et al. 2010; Kemper et al. 2011). By this time, assisted by data philanthropy, organizations like Amnesty International had begun to rent their own satellite time to sense evidence of human rights abuse remotely in places, like Darfur, judged unsafe to put staff on the ground (Prins 2008; Sulik & Edwards 2010). Moreover, by cutting staff travel and analysis time, the automation of feature extraction was lowering the access threshold to remote sensing for small and medium-sized NGOs. One geospatial scientist summarized the affirmatory powers of satellite recoupment as follows: 'It is a cheap data source with regular recording and can access and document large areas from "no-go" zones continuously and consistently, and thus provide image proof of before-and-after situations that can be produced more accurately, faster and more convincingly to the public than reporting from sources on the ground' (Prins 2008).

The growing ability to sense, map and extract humanitarian features and data from space has developed dynamically with growing restriction on international humanitarian access (Collinson & Elhawary 2012). As discussed in chapter 7, by the end of the 2000s, the combination of risk aversion, the roll-out of restrictive security protocols and increasing ground friction had combined to produce the phenomenon of the fortified aid compound. As remote sensing grew in accuracy, timeliness and availability, it has also normalized both the absence of and a decreasing reliance on ground truth. Reflecting the impulse of technoscience to turn political negatives into positive technological challenges, access denial is more a technical problem to be overcome rather than a reflection of an emerging nexus of long-term political problems on the ground. The very success of geospatial technologies in terms of sidestepping terrestrial insecurity and political push-back in the course of producing new sense-making tools reinforces the discursive distance between the observer and the observed. When we have these new tools, given the dangers and hassle, why go there? While the loss of ground truth can be lamented, digital recoupment creates new angles and possibilities. Refugees, for example, can now be viewed ecologically as biohuman actors in complex landscapes of limited resources and challenging logistics. Remote sensing, as we shall see, is also argued to bring objectivity, whereas ground truth, while useful, is susceptible to human bias.

Seeing is believing

Free from the problems and financial cost of being on the ground, through satellite sensing one can visualize the behaviour of refugees both 'objectively' and 'ecologically'. Camps became living parts of their surrounding environments. Revealed by multi-spectral sensors operating at different electromagnetic frequencies, the camp and its inhabitants were transformed into an interactive, multi-levelled feature within a given human/nonhuman milieu (Bouchardy 1995). From the outset, the objectivity of the satellite image was a topic of celebration. Because satellites record 'what actually exists on the ground, nobody can argue that the information has been omitted or changed, as can be argued if potentially biased individuals or companies carry out field surveys in remote areas' (Bjorgo 2002). As a UNHCR spokesperson argued, by being able to show from above what's being done in difficult environments, 'we can really highlight the challenges we face on the ground and how we tackle them' (Batty 2008). On the back of the commercialization of geospatial technology, the objectivity of the image underpinned a new wave of NGO public advocacy.

In 2005, Google launched Google Earth as a free, easy-to-programme open mapping platform. As early as 2008, over 350 million copies of Google Earth had already been downloaded (MapAction 2008). In April 2007, in partnership with the human rights group US Holocaust Memorial Museum (USHMM), Google facilitated the Crisis in Darfur project to showcase a new 'Global Awareness' layer within the Google Earth platform. As a composite of US military and commercial satellite imagery compiled between 2004 and 2007, unlike a conventional map, the Crisis in Darfur layers allow anyone with access to a screen interface to zoom down from space to see refugee camps, tents and burnt villages close-up while accessing photographs, videos, personal testimonies and agency reports on the way (Parks 2009: 536). The following year, Google released Google Earth Outreach. This combination of free software and technical support was aimed at encouraging more aid agencies to produce and share through the Google platform similar multi-levelled visualizations of their aid operations. As part of the Outreach launch, for example, UNHCR created pilot layers highlighting its refugee programmes in Darfur/Chad, Iraq and Colombia (UNHCR 2008). Like the Crisis in Darfur initiative, moving downwards from a regional overview, the viewer is able to explore the lives of refugees in specific locations 'by clicking on exact locations in the refugee camps to see photos of the facilities, such as health clinics, schools, water taps and sanitation' (Batty 2008).

At the same time as these public advocacy mapping initiatives were unfolding, in June 2007, the human rights group Amnesty International (AI) began its satellite-based Eyes on Darfur initiative. The Enough campaign, endorsed by George Clooney, launched its similar Satellite Sentinel Project in 2010 (SSP 2012). The human rights use of satellite imagery followed the same humanitarian feature extraction processes that had been developing for several years – in this case, using burnt dwellings as a proxy for human rights abuse (Sulik & Edwards 2010). Both of these initiatives enjoyed technical support from geospatial scientists and such companies as ImageSat International, DigitalGlobe and GeoEye (AI 2007). Besides releasing findings to the major news networks, they also used online activists and social media for dissemination.

These advocacy projects were based on the proposition that 'seeing is believing' (see Parks 2009). The human rights use of before-and-after satellite photography is based on the purported objectivity of the image (Lavers et al. 2009). A common refrain at the time was that, enabled by corporate philanthropy, human suffering in hard-to-reach places was now being illuminated as never before. It was being brought directly to public attention not in words, but in pictures. People could no longer say that they didn't know. Google Earth's multimedia layers were argued to help viewers experience, and thus better intuit, what was really happening on the ground. It was even possible to play a couple of online games that simulated the difficulties of being a refugee (HRE 2009), or an on-the-spot news reporter (Radio Dabanga 2009). Through the 'veritable media industry' (Parks 2009: 540) that had emerged around the Darfur crisis, advocates saw themselves creating an informed public force for change without burdening people with the need to read anything, let alone subjecting anyone to the danger of actually going there (Dobbs 2008). As John Prendergast, the co-founder of the Enough Project, said, regarding the David versus Goliath struggle taking place in Darfur, 'Google Earth just gave David a stone for his slingshot' (quoted by Parks 2009: 537). Not only could this ammunition be freely gathered without needing access permission, it didn't risk people on the ground and, apart from mobilizing the public, it reminded authoritarian states that the world was now watching (Lavers et al. 2009).

With the advocacy breakthrough failing to materialize, within a couple of years all of these initiatives had fallen silent and their respective websites were no longer updated. The crisis on the ground in Darfur, however, has continued (Jaspars 2015). At the same time, the Western governments that had initially condemned the actions of the Sudanese government in Darfur have now recalibrated their positions. Indeed, the same Arab militia groups that were responsible for much of the rapine, deaths and displacement have now, apparently, been recruited on behalf of the EU by the same, once-condemned government to prevent African migrants using Darfur as a passage to Libya and thus Europe (Grinstead 2016). These comments are not targeted at the people or agencies that attempted to publicize the dreadful things that were happening in Darfur in the mid- to late 2000s. They are more a reminder that we need to look past the hype and fetishism (Scott-Smith 2013) that surrounds smart technology. If the medium is the message, rather than being enthralled by the images, we need to ask what the medium is saying and where it is taking us.

In chapter 5, it was argued that, as a modernist humanitarian technology, the refugee camp functioned to separate, or wall-off, the politics of forced migration from society. The camp allowed a separation of political and humanitarian identities. The ideological emptying of the camps of any progressive content during the 1980s and, ideally at least, the integration of refugees within the community, saw security expand to encompass the whole of society. We had entered a hyper-political age.

Around the time that remote sensing became commercially viable, there was a growing feeling that, rather than temporary structures, as securitized spaces refugee camps were becoming permanent features. Reflecting the hyper-political, from the early days of humanitarian satellite sensing, camps have been sensed ecologically in terms of their interconnections with their surrounding environments (Bouchardy 1995). Besides available work or cultivable land, this includes refugee access to things like water, firewood or urban services. Remote sensing also facilitates the visualization of camps logistically relative to road quality, natural obstacles like seasonal water courses and distances from major distribution centres, operational agencies or areas of political unrest. Satellite sensing amplifies early warning's operational demand for timeliness, as well as objectivity. At the start of the 2000s, in terms of the satellites available to the aid industry, the frequency of revisits ranged from every couple of days to one or two weeks. Depending on the system being used, extracted data could be delivered in as little as one or two days (Bjorgo 2002). With aerial surveillance approaching near real-time, changes in camp size, even abrupt ones, can now be monitored remotely. At the same time, regular overflights mean that the environmental impact of camps – for example, in terms of land or forestry degradation – can be longitudinally mapped over months or years.

Separated by a techno-discursive distance between the observer and the observed, remote sensing reveals objects on the ground and, through the digital traces left by their behaviour, it is possible to visualize the refugee as an objectively surveilled subject (Harris 2006). With the history and politics of forced migration consigned to the background, the biohuman essence of the refugee moves to the foreground. The hyper-political subject of containment becomes an ecological and relational calculus of risk, associated with water availability, soil conditions, energy reserves, nutritional needs, logistical networks, medical services, sanitation infrastructure and preparedness for future uncertainty. In terms of realizing this ecological biohuman essence, however, remote satellite sensing by itself can only go so far. Remote sensing had sidestepped the problem of ground friction and access denial while creating new, objective and timely sense-making tools. There was still something missing, however.

By the mid-2000s, for example, satellites could identify individual buildings and dwellings within a given refugee camp. The anticipated next step was providing 'a kind of "zip code" and address for each inhabitant and family [to facilitate] the management of the evolution of the camp and its population' (Bally et al. 2005: 39). Although even temporary structures could now be given a grid reference, the individual behaviour of the biohumans within them could not be recorded, authenticated or predicted. While the camp could be visualized ecologically, the individuals within it could not be known – or, therefore, governed. This would be addressed through the rapid spread of mobile telephony that was occurring as the Darfur crisis was happening. The arrival of Big Data would complete the cybernetic transition from knowledge to data that first started in the humanitarian field in the 1980s.

Crisis Informatics

Except among Luddites, following its US countercultural rehabilitation during the 1970s and 1980s, the perception of computer technology is now free of earlier New Left concerns that it was a tool of corporate control and alienation. By the 1990s, computers had emerged reborn as the new economy's provider and guarantor of freedom, choice and personal creativity (Turner 2006). Given their steady diffusion throughout society, it now seems remarkable that, as recently as the mid-2000s, when the term 'Big Data' first emerged, using the logs of geolocational mobile devices to study behavioural dynamics was still experimental (Eagle & Pentland 2006). This was around the time of the launch of Google Earth, and the Darfur crisis becoming one of its first Global Awareness layers. Much of the foundational research on data informatics, especially its potential for new business and advertising models, was anticipatory and happening ahead of the rapid expansion of broadband connectivity (O'Reilly 2005). From the outset, it was realized that, by the very nature of the medium, mobile telephony makes it 'an ideal vehicle to study both individuals and organisations' (Eagle & Pentland 2006: 225). Knowing the content of calls or messages was unnecessary since structure emerges from routine. Logging the when, where and who of usage over time was sufficient to reveal the 'large-scale dynamics of human behaviour' (2006: 263). Enthused by the promise of mobile geolocationality, the latter half of the 2000s was an optimistic and expansionary time for the new business ventures of Silicon Valley.[8]

When social media was still in its youth, its rapid emergence promised to rejuvenate democracy through new forms of direct engagement, participation and active citizenship. Connectivity was the new *socius* that, in appearing to decentralize power, was argued to be creating new opportunities for progressive expression (see Reid 2009). In June 2009, for example, as hundreds of thousands of Iranians demonstrated in Tehran's Freedom Square against the impending election of President Mahmoud Ahmadinejad, the opening shots of the 'Twitter Revolution' were optimistically welcomed (*Washington Times* 2009). This emancipatory techno-optimism was also being extended to disasters. Building on the increasing use of remote satellite sensing among aid agencies, the main interest was in the implications of the rapid spread of broadband and mobile telephony, including within the global South. Spurred by Hurricane Katrina in 2005 (Crutcher & Zook 2009), in the two or three years prior to the Haiti earthquake (2010), 'crisis informatics' as a distinct area of operational possibility emerged in Europe and the USA (Muhren & de Walle 2010; Palen et al. 2010). Evidence culled from the records of mobile usage among people caught in college shootings, floods, forest fires, infrastructural breakdowns or contagious outbreaks in the global North was already showing that the new information technologies could play a transformational role in the way society responds to mass emergencies. Indeed, one could start to reframe disasters radically: 'as a set of *socially-distributed information activities* that support powerful, parallel, socio-technical processing of problems in times of change and disruption. Good quality information and meta-information that indicates accuracy and trustworthiness is what people need to make local decisions, to gain situational awareness and build resilience in the face of threat' (emphasis added, Palen et al. 2010: 10).

At the same time that academics were exploring the potential of crisis informatics to leverage systematically the 'collective intelligence of the public' (2010: 10), the UN launched its anticipatory Global Pulse project (UNGP 2009: 13). Alerted by the openness and vulnerability of the global precariat to the recent effects of the financial crisis, Global Pulse brings together UN agencies and the private sector to develop new approaches to behavioural analysis through the medium of Big Data. Encouraged by the promise of crisis informatics, the diffusion of data science to the international plane was seen as constituting 'a genuine opportunity to bring powerful new tools to the fight against poverty, hunger and disease' (2009: 4).

It is important to emphasize that the repositioning of disasters as socially distributed information systems, which Global Pulse embraces and exemplifies, has several supporting ontological and epistemological elements. Initially called crisis informatics, the structural transformation in question lacks a commonly agreed name. While subsequently attracting labels like 'digital humanitarianism' (Conneally 2011; Meier 2015), 'disaster relief 2.0' (Crowley & Chan 2011), 'cyber-humanitarianism' (Duffield 2013) or 'humanitarian innovation' (Betts & Bloom 2014), in the main, authors or agencies usually just allude to the changing nature of humanitarianism in the age of Big Data, or something similar (UNOCHA 2013). However, since the end of the 2000s, the core assumptions and propositions that constitute this transformation have been more or less continually reproduced in many different fora and publications.

All of these key assumptions can be traced to the NGO-led fantastic invasion of the 1980s. Here they emerged in an anticipatory form. They were experimental tools in a terrestrial struggle to convert history and politics into information and data. With the arrival of global connectivity they ceased being anticipatory; they have arrived. They now cohere and work together as a closed post-humanist operational logic. As they are now mutually reinforcing, some of the core assumptions and supporting arguments are rehearsed below. Besides seeing these elements as interconnecting and mutually supporting, it is also useful to consider this operational logic as a business model in its own right.

Growing impact

In infrastructural terms, growing connectivity has a downside. Due to cascading and multiplier effects, in a rapidly urbanizing world, the increasing frequency of disasters is now widely accepted, as is the inevitability that they will impact more people (Pelosky 2002). Further, contributing to this vulnerability, at an international level, 'our global economic system may have become more prone to large and swift swings in the past few years' (UNGP 2009: 11). Because this inevitability is hedged with uncertainties, however, while one can be alert, it is impossible to be fully prepared. Moreover, it is important to note that this is occurring at a time of austerity. Reasonable people can no longer expect, or take for granted, that public emergency cover will keep pace with this growing exposure. Humanitarian innovation anticipates the challenge of surviving in landscapes where any social contract between individuals, corporations or states is absent. Rather than a problem as such, these landscapes create opportunities to become something better.

Enabling resilience

As a form of radical self-help, digital humanitarianism embraces the centrality of resilience. Mobile phone usage confirms what has long been suspected. Rather than waiting for assistance, disaster-affected people and communities are their own first responders.[9] In ecological terms, 'how quickly the damaged part of an ecosystem can repair itself depends on how many feedback loops it has to the non- (or less-) damaged parts of the ecosystem(s)' (Meier 2013a). For humans, information technology creates the feedback loops or platform between the affected and unaffected within and beyond disaster zones. It thus allows a natural process of self-recovery to take place. Although 'real time' does not always mean immediate, feedback loops have to work within a timescale that allows for meaningful action. Mobile telephony, the internet and other software platforms that can build trust and aid coordination can be leveraged by responders, as well as creating 'relationships and information sharing portals with civilians' (Giroux 2009: 24). For disaster resilience, rationalizing how unaffected areas, communities and people help the affected through the bridging and logistical capacities of new technology is a key proposition. Thus, timely information is of equal – if not greater – importance than food, water or shelter.

Homo inscius

Homo inscius, or the necessary ignorance of the neoliberal subject, haunts the pages of this book. Building on the growing existential remoteness of international aid workers following their risk-related retreat into fortified aid compounds, *Homo inscius* has become a central player in the way crisis informatics reconfigures the cognitive landscape of disaster zones. Sudden disasters create arenas of fluid and constantly changing stimuli and signals that, by their nature, subject human decision-making to the stress of operational immediacy. New sense-making tools are required to redress the loss of meaning resulting from unexpected changes that 'break the imaginary link between expectation and reality and force actors to revaluate what they are doing and where they should go' (Muhren & de Walle 2010: 30). Disasters become post-humanist sites of unmediated empirical factuality resulting from the removal of familiar points of reference while challenging subjects to choose quickly between contradictory signals and demands. Research on sense-making, however, has shown that, despite cognitive dissonance, people do act; they do not wait passively, and, moreover, they give credence to 'what other people think and understand, and they take into account other people's reactions when they act' (2010: 30). This applies to the disaster-affected as well as emergency responders. This human condition creates a need for bridging information support systems that remove the burden of deciding. Such systems provide a behavioural guide in times of flux by facilitating sense-making through the timely exchange of value-added information between those who need help and those who can provide it.

Distributed information systems

Global connectivity radically extends the limits of spatial separation between the unaffected and the affected within disaster zones. These now stretch beyond the global South to encompass digital humanitarians working on the other side of screen interfaces beyond the barriers created to restrict and contain global migration. Regarding the early discussion of the fortified aid compound, increased connectivity allows for the appearance of the hyper-bunker. Many times removed from the disaster zone, the hyper-bunker multiplies the mobility differential of digital humanitarians. While there are earlier precedents of volunteer disaster mapping, the Haiti earthquake in 2010 proved the feasibility of data-mining social media and SMS messaging to extract timely humanitarian intelligence. This included identifying areas in need, together with points of refuge, which could not have been easily replicated on the ground. At the same time, this was made accessible to responders and affected alike through sense-making dynamic maps and messaging services. Importantly, however, it also demonstrated that this platform service could be provided through crowdsourcing the data cleaning and embedding work to new networks of young and largely self-organizing Volunteer Technical Communities (VTCs) (see Burns 2014).[10] Through their dependence on private 'data philanthropy', such VTCs have emerged alongside the growing humanitarian profile of Silicon Valley corporations like Google, Twitter and Facebook, together with large telecommunication companies such as Vodafone and Orange.

Decentralizing power

Within the canon of progressive neoliberalism, the transformation of emergencies into socially distributed information systems has, according to Tim McNamara of the Open Knowledge Foundation, not only involved a technological shift 'but also the rapid decentralisation of power' (quoted by Meier 2012). This decentralization has empowered new humanitarian actors, such as the hyper-bunkered voluntary crisis mappers, as well as the disaster-affected themselves. Whereas aid agencies once made all the decisions, people now have the informational tools to say what they need. In this way, 'communities and individuals are determining how to help themselves and how they want to be helped by others, mobilizing local, national and sometimes global support to meet their needs' (UNOCHA 2013: 13):

> Anyone can create information and interact with other people's information with a basic mobile phone. Coupled with the opportunities of Big Data and GIS technology ... a true partnership is possible in which citizens, communities and humanitarian actors collect data from a wide array of sources; transform raw data, through analysis, into useful information; freely share information with one another; and act on that information to save lives and prevent suffering. (UNOCHA 2013: 24–5)

Already discussed in relation to the fantastic invasion, this purported digital decentralization of power continues the post-Fordist trend of flattening professional hierarchies in the name of greater public participation (see Negroponte 2003 [1975]). One could argue that the folding downwards of smart technology is more about the hyper-centralization of power, as in the form of platform capitalism (Srnicek 2016). Moreover, the claims for empowerment are bound up in legitimizing both a commercial bid for a stake in the humanitarian market and the remote technologies that are employed. Thus, the traditional model of humanitarian agency which is funded and based in the global North, and which collects, analyses, decides and delivers assistance to the South, 'is now out of date' (UNOCHA 2013: 16, 23). The techniques developed by ground-based humanitarian actors, such as trying to get 'as close to the affected area as possible, find out what was happening and transmit that information to their superiors, sometimes by hand, sometimes in person', are now effectively redundant (UNOCHA 2013: 23; also Meier 2012). As a way of cementing these breaks, the radicalism of progressive neoliberalism – its rush to the future – is reflected in its direct embrace of the new economy and its vision of an empowering partnership between commercial technoscience and the global precariat.

Conclusion

In April 2013, to the applause of the new humanitarian actors (Meier 2013b), UNOCHA published its *Humanitarianism in the Network Age* report, which provides a state-of-the-art overview of disaster informatics, and the Volunteer Technical Communities and private companies involved, as well as outlining a programme of development. Soon afterwards, the International Federation of Red Cross and Red Crescent Societies (IFRC) followed with its own World Disasters Report entitled *Focus on Technology and the Future of Humanitarian Intervention* (IFRC 2013). While anticipated for a while, these publications mark the coming of age of the crisis informatics that underpins post-humanitarianism. The reports interconnect and reproduce all the core operational assumptions outlined above. Remote sensing is seen as giving a new objectivity to disaster zones. The refugee camp, for example, can now be viewed in relationship to its environment, and refugees reinterpreted in terms of the opportunities and constraints revealed. At the same time, there is a consensus that disasters are becoming more complex and their impact greater. Regarding the latter, the rapid flux of events in a disaster also disrupts normal cognition. They are events that exemplify the challenged nature of *Homo inscius* – both as a subject requiring assistance and, more importantly, one that needs to be taught how to help itself. In this context, the development of crisis informatics appears fortuitous. Disaster-affected communities and those responding to such events have been rediscovered as distributed information systems. Not only has this allowed new humanitarian actors to appear, it is argued to have decentralized power while making new sense-making tools possible. In taking this analysis further, the next chapter considers the set of practices known as humanitarian innovation. This initiative addresses the infrastructural crisis through the development of personalized objects and self-acting technologies that, together with crisis informatics, are capable of supporting individuated forms of nomadic off-grid existence.

Chapter 11

LIVING WILD

It was argued in chapter 9 that smart technology levels downwards. As it attentively accommodates and adjusts to inequalities and immobilities encountered, it encodes and reproduces them. Previous chapters have emphasized the absence or dereliction of a universal fixed infrastructural grid as a characteristic of the global South. For technoscience, this does not present itself as a political and economic problem but as a technical challenge – that is, as something to be worked around, side-stepped or designed away. While crisis informatics provides the means and rationale for information feedback, it does not address the absence of a fixed-grid. This chapter looks at the role of commercially supplied humanitarian objects and smart technologies designed to work off-grid in challenging environments. These range from cash transfer and solar lighting to therapeutic foods and personal water filtration systems. As a field of interest, it is usually known as humanitarian innovation or humanitarian design (Nussbaum 2010; Betts & Bloom 2014). These terms are used here interchangeably.

Rather than seeing these objects simply as a substitute for a fixed-grid, this chapter is concerned with the shared logic that connects them. Together, they constitute a surveillant apparatus that allows movement while encouraging disaggregated and personalized forms of post-social survival. As 'humanitarian' technologies, they have absorbed the indignation that drove the direct humanitarian action of the 1980s. Liberalism's enduring disgust at want in the world has now been transformed into commercially supplied self-acting objects that work remotely. The fantastic invasion's notions of the 'project' and the 'community' have likewise been absorbed into these technologies. While communities were once encouraged to become self-managing, we now have communities of 'users' who are permanently enrolled in the continuous prototyping of the technologies that govern them. The chapter concludes by considering how the success of humanitarian innovation rests upon a narrow empiricism regarding what counts as evidence of success; and how this narrowing relates to the changing nature of ethics.

Humanitarian Design

Since the mid-2000s and the increasing salience of Bottom of the Pyramid (BOP) economics and notions of inclusive capitalism, the corporate marketization of poverty has been evident (Schwittay 2011). One register of this has been a rapid growth of commodities, objects and financial products aimed specifically at the precariat (Cross & Street 2009). Reflecting this development, humanitarian innovation is defined as drawing 'upon concepts from the private sector to adapt and improve the humanitarian system'. This includes leveraging and combining the new partnerships and technologies of the former with 'the ideas and coping capacities of crisis-affected people'. It encourages 'adaptation and improvement through finding and scaling solutions to problems, in the form of products, processes and business models' (Betts & Bloom 2014: 5). Tom Scott-Smith has called this historic combination of humanitarianism, technology and the market 'humanitarian neophilia' (Scott-Smith 2016).

In seeking to distinguish this movement from past aid efforts, humanitarian innovation's claim for legitimacy rests, in part, upon an open acknowledgement of failure. Or, at least, an acceptance that, after sixty years of terrestrial development, perhaps a billion or more people in the global South still lack what would be classed as bare essentials in the North: access to piped water, stable electricity, proper waste disposal, adequate housing, comprehensive schooling, professional health care and regular financial services. Levelling-up or reconstructing a fixed-grid, however, hasn't been a serious international objective since the beginning of austerity at the end of the 1970s. This aversion has deepened with the growth of slums, the increasing infrastructural damage caused by natural disasters and the massive urbicidal destruction unleashed in the Middle East.

When faced with the infrastructural gap, the new economy prefers personal optimization, behavioural adjustment, gentrification, segregation, speed differentials and design solutions. Be they climate change, insecurity, pollution, chronic poverty or entrenched interests, technoscience ducks underlying political and economic causes in favour of work-arounds, fixes, adaptations and short-term wins (Joseph 2016). It embodies the reduced horizons of what Orit Halpern has called 'hopeful' resilience (Halpern 2017). Smart technology folds downwards and normalizes the inequalities it encounters. For humanitarian innovation, the infrastructural gap and absence of a social state constitute the design challenges it faces. From such a low starting point, any improvement is a win. Peter Redfield quotes a manager from the Danish company that makes the LifeStraw, a personal water filter, who candidly confides: 'Let's be honest, we're not getting a municipal water system in rural Kenya anytime in the near future' (Redfield 2015: 16). Humanitarian innovation flags global precarity as a sign of self-evident failure, yet, at the same time, it is the terrain on which it stakes its own future success.

In practice, the objects and technologies of humanitarian innovation are unevenly distributed throughout the global South. We are concerned to bring them together here in an ideal sense, and to look for logic or commonalities of action that build on the potential of data informatics and the cognitive turn. While its stakeholders ignore or exclude certain things – not least, meaningful social change – this operational logic allows a post-social capitalism to experiment and innovate within its own limits and values as it attempts to make real a future that, at the moment, is still in formation. In order to explore this logic, a few of the new humanitarian technologies are outlined first.

Cash transfer

Following on the heels of mobile telephony, an important innovation has been the celebrated spread of mobile banking among the precariat in countries where they had been previously excluded from financial markets (Iazzolino 2015). Emerging alongside this development, directed towards the chronically poor, cash transfer programmes are now widespread. Often run by a combination of banks, telecommunication companies and NGOs, as an alternative to public provision or in its absence, these programmes typically involve the periodic transfer of small amounts of money enabling the purchase of commercially supplied services or food (Donovan 2013; Lavinas 2013; ODI 2015). While still subordinate in terms of coverage, cash transfer is also becoming an increasingly favoured alternative to international food aid (World Bank 2016a). As discussed in more detail below, such programmes usually involve biometric registration.

Water

In the absence of a fixed water supply, including wells and hand pumps, a number of companies are producing personal filtration systems capable of purifying contaminated surface water. Among the better known is the LifeStraw (Redfield 2015). While larger units are available, the personal version is a portable ceramic filter around 20 centimetres in length. Having no moving parts and not dependent on electricity, it works by the direct action of sucking. Each device is claimed to be able to deliver thousands of litres of 99.9 per cent pure water before needing replacement.

Emergency shelter

There is a big interest in emergency shelter, with many examples on the market or in design studios. As an innovation challenge, it must demonstrate usability and affordability (Fredriksen 2014). Better Shelter, for example, has been produced by a Swedish design company in collaboration with IKEA and UNHCR since 2010. Complete with a solar panel in the roof which can provide light and charge mobile phones, Better Shelter is a single-room modular structure built from a steel frame and polyolefin panels. The company's web page explains how the structure, besides offering more safety and dignity than a tent, has been designed to resemble outwardly a home with windows and a lockable door. It is also easy to dismantle and transport.[1]

Energy

Improvements in solar-panel and battery technology have catalysed a huge growth of a wide range of portable lighting and device-charging solutions (Cross 2013). The advantages of solar power are typically set against the expense, danger and health risks of kerosene lamps. For example, the social enterprise Solar Aid markets solar lighting through school networks and local enterprises. Solar Aid's website emphasizes how clean and cheap solar lighting can boost child learning and feelings of wellbeing, and constitutes a 'first step on an energy ladder to full electrification.'[2]

Sanitation

Various technologies now exist to overcome the absence of a fixed sanitation system, or even the need to dig pit latrines. Such waterless toilet technologies usually involve a system for composting human waste (Robins 2014). The Peepoo bag, for example, is designed for personal use. It consists of a biodegradable plastic sac that has been chemically treated. Besides containing smells and disease, they are marketed as a useful source of fertilizer.[3]

Therapeutic foods

Given their cheapness and international marketization, instant noodles are currently seen as an updated proletarian food for the global precariat (Errington et al. 2012). At the same time, commercially supplied therapeutic foods that address child undernutrition now have an established presence. One of the more well-known is Plumpy'Nut.[4] This is a peanut paste containing added nutrients and vitamins. It can be administered in the home, circumventing the need for mother and child to attend a feeding centre (Scott-Smith 2013).

Self-help apps

There is an expanding range of e-learning and self-help apps adapted for mobile phones. Facebook's Free Basics, for example, offers access to basic websites for news, job postings, health, education and communication tools (Gillula & Malcolm 2015).[5] Health apps, for instance, take the form of self-diagnostic tools.

Drone delivery

With load potentials of over a ton, and rough landing and take-off capabilities, drone range and capacity are growing.[6] With air-drop capability, one can anticipate that this technology will become increasingly encountered in challenging environments (Crowe 2013).

Individuation

The lack of civil registration among the precariat, including the recording of births and deaths, has long been regarded as a problem for health management (Setel et al. 2007). That up to 1.5 billion people currently lack legal proof of identity is also argued to weaken economic participation (World Bank 2017). Since 2001, UNHCR has been experimenting with the biometric registration of refugees. This usually involves either fingerprinting or taking iris scans. In 2010, such registration was announced as official UNHCR policy (Jacobsen 2015). The European refugee crisis, intensified by the conflict in Syria, has seen the biometric registration of refugees rapidly expand (Slim 2015; FindBiometrics 2016). Biometric technologies are now widespread throughout the global South. They have seamlessly become a precondition of mobile banking, cash transfer programmes, aid entitlement and citizenship claims. In Africa alone, millions of people have been fingerprinted or scanned within the last decade (Hosein & Nyst 2013). Cisco is currently experimenting with blockchain technology, associated with the development of the Bitcoin, as a means of developing fluid and transposable forms of refugee identification. Its recently unveiled Freedom-As-A-Service proposal involves adapting this technology not only to authenticate identity but also to measure, validate and record the work and service contributions of chain members (Morrow 2016). The aim is to encourage a form of humanitarian gig economy whereby refugees barter, lend, rent, trade or swap skills and resources among themselves.

Indignant Objects

Reflecting Bruno Latour's approval of how the sentiments of design have trumped those of revolution (Latour 2008), post-humanitarians have enthusiastically embraced the design principle. As more architects, artists and engineers are attracted to the challenge, there have probably been more exhibitions, installations and curated events devoted to humanitarian design over the last two or three years than over the past decade or more.[7] When objects become disarticulated post-humanist things that have a separate life in the wild beyond human experience, the more amenable they are to the design principle. There is a related tendency among practitioners to fetishize the technologies of humanitarian innovation. They are often imbued with miraculous developmental, curative or rehabilitation capacities. As Tom Scott-Smith suggests, it is important to try to 'see past the possibilities' of such technologies and objects (Scott-Smith 2013: 198). Together with crisis informatics and cognitive development, these post-humanitarian technologies and objects offer a glimpse of what 21st-century global welfare is beginning to look like. The logic that connects these technologies differs from the insurance-based normative biopolitics of the welfare state. It is also distinct from the more recent community-based development of the fantastic invasion. Before these differences are examined, however, it is necessary to establish the status and action-orientation of this logic.

Chapter 5 argued that some of the energy of the May '68 critique was recouped in the direct humanitarian action of the 1980s. Seeking solace from revolution's failure, this displacement translated into a willingness to spend time on the ground in a direct struggle against exclusion. While transmuted, an echo of this recoupment still resonates in the humanitarian object. We find, for example, the same indignation over absence and failure. This time – rather than celebrities – it is now architects, engineers, design companies and social enterprises that are demanding: 'Why is there no functioning water system? Why do the roads remain poor? Does anyone remove the trash?' (Redfield 2015: 17). For humanitarian innovation, however, indignation does not appear as direct action on the ground; it transmutes into the *direct action of the object*. It is the object's ability to work remotely, automatically and in the absence of external support or detailed knowledge that is praised and designed for. It is the object that is now indignant; and the fewer working parts, the better.

This is not a case of 'technological determinism'. For example, a history of imperialism can be written in terms of the steamships, undersea telegraph cables, maxim guns and quinine that made it possible. From this perspective, technology is a prosthesis acting as a force-multiplier for the imperial impulse. Human agency, however, remains in the loop. The humanitarian object is different. Its self-acting nature suggests that human agency has been transferred and absorbed. Once created, as explored further below, the relationship of humans to smart devices is one of servility, slavery and, ultimately, complete irrelevance (Spencer 2016b). To be able to exert this power, however, smart devices have to work together as a system of *interconnected* devices and technologies – indeed, as an internet of things.

Such objects require, aside from the technical challenge of interoperability, a shared design logic that anticipates and makes connectivity possible. While anticipatory, this logic is the system in formation.

Connected logic

Compared to social insurance-based welfare systems, one of the more obvious differences with the post-humanitarian technologies described above is that they are commercially supplied and work through market principles (Nussbaum 2010; Johnson 2011; Lavinas 2013). Rather than operating at the aggregate level of a statistical population, or within a given community of people, they are personalized or individuated in some way. This either concerns payments or transfers to an identified individual, or involves the personal use or consumption of energy, resources or services. In some cases, self-administration or self-learning is involved. Against a background of global containment, they also allow for movement or nomadism. The objects or services involved are either portable or can be accessed through mobile devices or biometric terminals. They can be carried or transported from place to place, consumed as a web-service, or physically received through identity-verified distribution points. When taken together, these technologies 'function as a substitute micro-infrastructure, one divorced from any project of extending a grid of urban services' (Redfield 2015: 16). They substitute an expensive fixed-grid while allowing surveilled movement that can, as a result of its recording, be logistically optimized.

Another important feature of these technologies is that they blur the interface between economy and disaster. They work equally for both the chronically poor and disaster-affected groups. Such adaptability thus addresses the permanent emergency of precarity while creating a market that is potentially commercially viable. They also lend themselves to remote management, user surveillance, product feedback and periodic updating. This surveillant apparatus is returned to below, where the relationship between humanitarian innovation and the project form is discussed further.

These technologies are widely celebrated as offering innovative, evidence-based responses to the challenging environments of the global South. They also reproduce key elements of the spirit of progressive neoliberalism, such as privileging the sentiments of personal freedom, choice and equity. In some cases, for example, through cash transfer or therapeutic feeding programmes, the targeting of women and vulnerable groups is possible. By emphasizing personal control, humanitarian innovation continues the work of the fantastic invasion in providing alternatives to patriarchy and the negative hold of tradition. Despite being associated historically with the registration of criminals (Agamben 2008), the hyper-personalization made possible by biometrics has been welcomed (Slim 2015; World Bank 2017). Biometrics are celebrated for sidestepping the lack of traditional forms of ID, such as birth certificates or passports, which is often a feature of precarity in the global South (Duffield et al. 2008: 25–6). Their use in cash transfer systems or in certifying aid entitlements is valued for minimizing fraud (Donovan 2013). Likewise, the portability of many humanitarian objects is also regarded as reducing opportunities for theft.

In terms of their mode of operation, post-humanitarian technologies not only provide an alternative to a fixed-grid, they lower logistical costs, reduce staff requirements and minimize professional involvement. With the onus on the individual to purchase food or services directly from the marketplace, the trend for voucher or cash transfer to replace food aid is welcomed as simplifying the logistics of conventional aid operations (World Bank 2016a). The wide adoption of Plumpy'Nut, for example, becomes 'revolutionary because it took therapeutic treatments out of the clinic and into the community' (Scott-Smith 2013: 916). As a means of addressing undernutrition, traditional feeding centres are regarded as disempowering of women by taking them away from the home and work. With each sachet of Plumpy'Nut constituting an individual portion and two or three being required per day, mothers are now able to self-administer this treatment at home until the child is discharged from the programme. This has reduced the need for professional staff and feeding centres which were 'expensive to maintain [and] costly in time, space and resources' (2013: 916). Humanitarian innovation thus continues the process of dissolving professional hierarchies in the name of streamlining and popular participation. As discussed in chapter 2, this liquidation is a signature move in the transition from Fordism to post-Fordism.

The streamlining of access to off-grid services and support reduces immediate costs. Not least, it dispenses with the jobs and resources that were necessary to make a fixed-grid function. In this respect, an important but neglected factor is the fate of local aid workers. Locally recruited aid workers, by far the majority in any aid programme, are widely impacted by these technologies. Since the 1980s, many became expert in the evaluation and management methodologies of terrestrial aid, including the feeding centres mentioned above. These skills are now rapidly becoming redundant (Jaspars 2015: 180). Apart from data-input, post-humanitarianism creates few – if any – jobs on the ground.

The reader may object that reconstructing a fixed-grid or creating jobs is not the role of humanitarian assistance. This objection, however, highlights the ambiguity of these 'emergency' technologies. While marketed with this label or emphasizing their humanitarian intent, they are appearing in contexts where, by common admission, there is little prospect of an alternative. Vestergaard Frandsen, the company making LifeStraw, addresses this ambiguity by using 'the terms humanitarianism and development interchangeably ... We're a company that says let's do what we can' (quoted by Redfield 2015: 16). Having no vision of an alternative future, the brittleness of this 'can do' practitioner ethic becomes clear when the situation is presented differently. Imagine an advert where the smiling children frequently used in the company's publicity material are somehow depicted as sucking contaminated surface water through a ceramic straw for the rest of their lives. Again, it could be objected that this is a harsh comment since social change is not the responsibility of private companies. As already mentioned, however, smart objects and technologies are not neutral; human reason and agency have been transferred and absorbed by them. Not only is there an acceptance of the status quo, connectivity and the cybernetic *episteme* work against independent and autonomous change. Independent and offline action contradicts the surveillant logic of command and control.

User communities

It was argued above that the indignation that drove humanitarian direct action has been transferred to the remote self-acting properties of the smart device. This absorption has also changed the action-orientation and meaning of the 'project' and the 'community'. As we saw in chapter 6, both concepts were central to the 1980s livelihood regime. The project form, it was argued, was anticipatory regarding the idealized post-social work and career structures then emerging in the global North. At the same time, it was also an intermediate step in the expansion of capitalism beyond the colonial wage relation, to absorb the then-autonomous areas of community mutuality and social reproduction. This was a crucial step in the emergence of a global precariat. At the time of the fantastic invasion, the 'community' usually comprised an operationally manageable number of people sharing some ascribed identity or interest. Often educational or disciplinary in nature, by preparing or empowering communities, projects were the practical means through which markets would be made to work for the poor.

Some objects of humanitarian design – solar-powered lamps or personal water filtration systems, for example – bear a passing resemblance to the utilitarian tools and artefacts encouraged by the Intermediate Technology (IT) movement initiated by the work of E. F. Schumacher (1974). Besides fuel-efficient stoves, typical IT artefacts included improved village wells or hand pumps to retrieve ground water (Redfield 2015). As intermediate steps in the developmental process, an NGO well or hand-pump project would typically enrol the local community in siting these features, and in providing labour for their installation and long-term maintenance. Intermediate technologies were ways of constituting the community as a group of actors. At the same time, these artefacts were often a precursor for associated disciplinary or educational projects. While equipment and materials were supplied upfront by the NGO, to encourage community ownership, cost-retrieval was frequently pursued through collective repayment over time. Additionally, the installation of pumps and wells provided opportunities for educational projects, often targeted at women, on hygiene and cleanliness.

In practice, however, once installed, training completed or costs recovered, pumps and wells were often forgotten by the initiating aid agency. They became ticked boxes as the NGO moved on to new successes. Eventually, many such projects would become fouled or fall into disuse due to the absence of spare parts. Once technology ceases to be intermediate and becomes smart, however, no such forgetting is possible.

Although imagined 'communities' still exist, their action-orientation is different. At a rhetorical level, community is now a homely dimension within the techno-pastoral aesthetic which valorizes the diffuse reproductive networks of the precariat (see chapter 9). More specifically, however, like the indignation that drove direct humanitarian action, the project form has been transposed and absorbed into the technologies and objects of humanitarian innovation itself. We now have communities of *users* whose personal data functions continually to refine and update the smart devices they use to navigate a post-social wild. Of course, this is a thoroughly artificial or ecological wild – evacuated of state welfare responsibility, it is a digitally embedded, pervasively surveyed and remotely sensed wild (Gabrys 2016). Within such environments, attentive technologies are today's never-ending projects. Reflecting the design ethos of attentiveness and care (Latour 2008), the advocates of humanitarian innovation emphasize 'the importance of test and real-world feedback' (Redfield 2015: 16) to allow the reflexive adjustment and continuous improvement of their products. Humanitarian innovation is attentive in that it rests on the back-and-forth of continuous prototyping.

Intermediate technology was a transient means of realizing an imagined community. The labour of installing the pump or digging the well was limited in time and space. Once installed, they were left with the community to use or forget. In contrast, the new user communities of smart technology are permanently enrolled. Their task of producing data for the reflexive adjustment of commercial objects is continuous and never-ending. The use of the LifeStraw, for example, is monitored through mobile apps and repair centres (2015: 10–11). The transience of intermediate technology was not an oversight. Whether workable or not, it was part of its imagined role in kick-starting a self-help model of development. While small-scale and dispersed, things like wells and pumps could look towards the levelling-up promised by a fixed-grid. By virtue of their positioning as 'intermediate', one could read in such technologies a potentially larger collective future.

Associated with the flat ontologies and immediacy of post-humanism, humanitarian innovation negates such a future. In the transient enactment of its installation and operation, intermediate technology realized the community as a collective of potentially self-directed actors. As a collective of users, however, it is the daily interactions of the community that now serve to realize smart technology. Indeed, as data is freely extracted, the relationship between users and smart technology has all the feel of servility, if not slavery (LeBaron & Ayers 2013; Stiegler 2016). Reflexive adjustment involves the continuous surveillance, recording and mnemotechnic absorption of the hard choices of the precariat. It is on this exponentially expanding body of recorded data gleaned from transactions, app usage, updates, medical reports and customer surveys that cognitive development and humanitarian innovation intersect and blur together in their designs for commercially scalable guidance and support. Besides continuous community enrolment in prototyping, as we shall see below, many of these products are involved in green and social marketing strategies that must constantly engage a distracted Western public.

Measuring Success

In terms of its direct aims, humanitarian innovation undoubtedly works. The technologies and objects reviewed above do make a difference in the immediate lives of the precariat. Regarding therapeutic foods, for example, scientific studies show that they are effective in reducing child malnutrition (Briend & Collins 2010). Likewise, solar-powered lanterns work every time the switch is depressed. One could say that these devices are 100 per cent effective in transforming darkness into light. A LifeStraw will deliver thousands of litres of 99.9 per cent pure water during the course of its serviceable life. Since a person needs 2 litres a day for optimal functioning, at $20 a unit this is a bargain. However, while these technologies indisputably work and are adapted to off-grid environments, they can be morally and socially ambiguous. They are not easy to love, and prove hard 'to simply embrace or fully reject' (Redfield 2015: 2). In order to appreciate them properly, a certain lowering and redirecting of attention is required.

The 'evidence based approach' (DFID 2012) that has emerged to showcase the success of post-humanitarianism works by black-boxing the history and structural inequalities of the wider society where practitioners operate. What counts as evidence relates only to the direct results of a particular intervention. Any surrounding social or political noise is excluded. While smart devices rely on continuous audit and feedback, absent from this attentiveness 'is a middle level of abstraction between situated actors and far-flung networks' (Redfield 2015: 16). In other words, the narrow empiricism involved excludes the commons – or the contested ground of theory and politics that lies between reality and the world. In a word, it negates capitalism. The evidence-based approach, however, goes further than just black-boxing. Since an important measure of success is the extent to which such technologies are commercially scalable (USAID 2013), solutions that do not challenge the status quo or cause ground friction tend to move into the foreground.

In order to claim success, the evidence-based approach medicalizes social problems. Plumpy'Nut, for example, is marketed as a medical treatment, while LifeStraw lays claim to the purity of its product. Where this cannot be readily done, problems are the result of sub-optimal behaviour and thus open to cognitive therapy. In the case of food security, for example, there have been significant changes in how human nutrition is taught within the academy since the 1980s (Jaspars 2015: 12–13). Reflecting similar changes discussed in this book, food security is no longer seen structurally as a causal outcome of the interplay of asymmetries of mobility, information and power. Neither is it addressed through livelihood support or direct humanitarian action, as during the fantastic invasion. It has been localized, personalized and, importantly, medicalized.

Having lost its social dimension, for the past decade or more, food aid has been understood in relation to the medical importance of nutritional intake, individual capacity and recipient knowledge. Nutrition is receptive to cognitive messaging and embraces the culture and practice of food preparation and hygiene. At the same time, it establishes the dietary intake necessary to support pregnancy and child development. In other words, it is an important factor in the disaggregated biopolitics of the post-social. As discussed further in the next chapter, in World Bank terms, good nutritional practice reduces the cognitive tax that the everyday hassle of survival imposes on the precariat. Undernutrition causes stunting, wasting and impaired cognitive development. As a medical condition, it results from deficiencies in energy, protein and essential vitamins and minerals. Maternal and child undernutrition, for example, is the underlying technical cause of 3.5 million deaths, 35 per cent of the disease burden in children under five years old, and 11 per cent of the disability-adjusted life-years (DALYs) lost globally (Ruel & Alderman 2013: 543).

The medicalization of such social attrition has occurred at the same time as the disappearance of social reform from the food security agenda (Jaspars 2015: 54–5). Beginning in 2008, for example, an influential series of papers published in the *Lancet* drew attention to how the disease burden of undernutrition among poor mothers and children could be reduced by targeted interventions that focused on making the best use of available household food (Black et al. 2008). Although it acknowledged that 'addressing general deprivation and inequality would result in substantial reductions in undernutrition and should be a global priority, *major reductions in undernutrition can also be made through programmatic health and nutrition interventions*' (emphasis added, 2008: 243). Medicalized interventions focus on meeting the dietary and micro-nutrient requirements of social reproduction through improving women's health, together with optimizing their behaviour regarding breastfeeding, hygiene and food preparation. Besides parental training, nutrition-sensitive programming is able to call upon the smart technologies and objects described above – for example, cognitive messaging, cash transfers and therapeutic foodstuffs, together with improvements in the off-grid supply of water, sanitation, shelter and lighting (Ruel & Alderman 2013).

These *Lancet* articles provide the evidence that smart medicalized interventions can improve social reproduction among the precariat without the need to address 'general deprivation and inequality'. With the first article appearing at the time of the 2008 global financial crisis, this series justifies the status quo while helping to provide an ethical basis for market-based humanitarian innovation. At the same time, however, we can see in these moves the already-discussed techno-pastoral aesthetic – that curious appearance of change, as in the success of medicalized interventions, while things remain essentially the same. Precarity and slums still exist, but it's now a wired precarity and a smart slum. In relation to the humanitarian objects described here, the techno-pastoral is evident in the publicity photographs used to advertise them. Invariably, one sees happy people against a backdrop of poverty, want or tradition. One assumes they are smiling because of the self-acting objects they are holding or using. At the heart of the techno-pastoral aesthetic is an absence. It's the black-boxed 'middle level of abstraction' between reality and the world. With no way of envisioning a better future, smart technologies that fold downwards into poverty keep reinventing its timeless presence.

Chapter 5 described how, during the fantastic invasion of the 1980s, the appearance of complexity-thinking saw the disappearance of history, society and causation. Things like famine, for example, rather than freeing their victims for new forms of capitalist incorporation and exploitation, seemingly exposed them to a naturalized 'hostile environment'. It was the beginning of the ontopolitics of the

Anthropocene (Chandler 2018). Three decades later, the shift towards medicalization continues the same manoeuvre. Because capitalism, or the content of the 'middle level of abstraction', is absent, the risk factors affecting the nutrition of the poor are generalized in terms of its epiphenomena of climate change, the volatility of global markets, the growing impact of natural disasters and the irrationality of terrorism and conflict (WEF 2010: 1376). The replacement of an absent middle by an uncertain environment suggests that the imbuing of objects with direct action and the localization of evidence are also bound up with a new ethical sensibility.

Ethical irony

The indignation of the May '68 movement was recouped during the 1970s and 1980s as direct humanitarian action. Humanitarian innovation's acceptance of the status quo marks the end of the line for this *sans-frontières* radicalism. This attenuation can be seen in relation to the changing nature of humanitarian ethics. The narrow empiricism that supports humanitarian design's claim to success has an associated regime of sentiment. The absent middle of the techno-pastoral aesthetic is its black-boxing of capitalism. References to the concrete historical, social or political determinants of precarity are now backgrounded in favour of anthropocentric environmental concerns. The *solidarity* that was central to New Left internationalism has been transformed into what Lilie Chouliaraki has called an ethics of *irony* (Chouliaraki 2013). Ethics has followed the narrow empiricism of success – indeed, post-humanism generally – to focus on what is local, immediate and inner to the self.

While earlier forms of solidarity were informed by pity, and acted on the basis of a common humanity and expected nothing in return, the new ethics reflects the scepticism that now exists regarding such a 'common humanity'. For Chouliaraki, regarding the suffering of others, the Western public has become an *ironic spectator* – it now has a certain detachment and self-conscious suspicion with regard to the truth claims being made. The negation of theory has left a disjunction between what is said and what exists – that is, between reality and the world. Without history, theory and politics, there is nothing anymore to hold what is said and what exists together. Certainly, such a task is beyond the ability of *Homo inscius*. Not only is the existence of a common humanity in doubt, so are all claims to truth. It is no accident that, for example, the Collins *Dictionary* declared 'fake news' its 2017 word of the year.[8]

Given the uncertainty as to truth, the ethics of irony uses the suffering of others as an invitation to consider and explore one's own inner condition via the emergence 'of a self-oriented morality, where doing good to others is about "how I feel" and, must, therefore, be rewarded by minor gratifications to the self – the new emotionality of the quiz, the confessions of our favourite celebrity, the thrill of the rock concert and Twitter journalism being only some of its manifestations' (Chouliaraki 2013: 3–4).

We can add that, in a post-truth world, science is the only vector having the potential credibility to attract a consensus. If the ironic spectator is to be moved, the evidence-based support for the immediate effect of the self-acting humanitarian object is important. When truth is uncertain, the constant surveillance, medicalization and prototyping are reassuring. Humanitarian innovation satisfies the demand to be seen to 'make a difference'.

The appeal of humanitarian design to the ironic spectator is further encouraged by the grounding of the former within the trope of environmental responsibility. The immediacy of impact is reinforced by the design appeal of words like 'reusable', 'sustainable', 'degradable', 'recyclable', 'neutral', 'responsible' and so on. Ideally, humanitarian objects inhabit the green economy. In the case of LifeStraw, for example, carbon-offset trading in Kenya provides the financial subsidy necessary to make the product commercially viable. The argument is that the more Kenyans use a LifeStraw, the less firewood is burnt to boil and purify water, and thus the less local carbon is produced. Besides data necessary for continuous prototyping, this financial model also requires constant user monitoring to make a case for offsetting carbon production elsewhere in the world (Redfield 2015: 10–11). Carbon-offset trading, which mixes ethics and finance at a global level, reinforces the aesthetic of immediacy with a claim of environmental responsibility. One could unpick the contradictions in this position – for example, how it works against the fixed-grid solution to water and sanitation that people prefer (Cross & Street 2009; Robins 2014). It is important, however, not to lose sight of how such a presentation appeals to the ironic spectator.

Many of the humanitarian objects described here have a secondary market in the global North. Since they are designed to function off-grid, they are widely found as camping accessories and wilderness survival gear. They enable ironic spectators, if you will, to connect with a natural wild as a recreational pursuit. Since the same material objects are involved, we can see in this use an interesting commentary on their role in the global South. However, this is not so much the unsettling moral contrast between a leisure activity and the hard choices of precarity. More important is the logic that connects them. They reflect the opposite poles of the new individuated biopolitics of the biohuman that has all but replaced the normative social-insurance model of the welfare state.

Rather than a dependence on collective institutions, in anticipating a post-social world the demand is for personal responsibility in all matters. In health, for example, issues of personal lifestyle, regular exercise and bodily fitness have moved to the foreground. Intolerance towards sub-optimal behaviour is growing. As wilderness survival gear, the humanitarian object helps to manifest the desired aesthetic of this biopolitics. It showcases the agility and ability to propel and maintain an independent bodily fitness. In the global South, we have an attenuated version of this freedom. Besides attempting to optimize behaviour, the humanitarian object is part of a medicalized biopolitical apparatus that delivers the minimum essential inputs required for off-grid physical and mental survival. Coupled with the smart messaging of cognitive development, it seeks to maintain the basic social reproduction of the precariat. While different and separated by a significant life-chance divide, the biopolitics of personal fitness and the medicalization of humanitarian intervention interconnect. One is an elite ideal, and the other a global reality. Together, they anticipate the contours of the post-social world now in formation.

Conclusion

The operational logic that informs humanitarian innovation is an ontology and methodology of closure. The only points of censure or barriers that it admits – things like data privacy or ethical considerations, for example – are factors that are already internal to the system itself. Even if they could be addressed, rather than fundamentally changing things, the outcome would be to deepen and extend the system. As a totalizing *episteme*, cybernetics was singular in being able to formulate its vision of a connected world in advance of the machines, data-parks and algorithms necessary for its realization (Halpern 2014c). As earlier chapters have demonstrated, several decades of anticipatory ground work preceded the computational turn. The rise of empiricism and behaviourism in the academy, together with the emergence of complexity-thinking and early warning systems within the aid industry, were bound up in the transition from Fordism to network capitalism. Direct action has been transmuted into a system of self-acting objects no longer requiring a fixed-grid to operate, and to which human reason has been transferred and subordinated. This is occurring at a time when resistance, political push-back and ground friction are growing. When critique is needed more than ever, progressive forces are incorporated within the new economy, where they provide the drive and creative energy to produce a techno-pastoral ethic where things change but remain the same.

Rather than marking a point of departure and the start of a new era, the computational turn appears more like a long-anticipated arrival. Distanced from direct engagement, rejecting a need for meaningful social change, and denied critique by a post-humanism that accepts no world beyond one's immediate enfolding environment, the present has all the feel of a time of ontological and technological closure.

Chapter 12

CONCLUSION: AUTOMATING PRECARITY

We have moved from an age that valued reason and human agency to a world where their stock has depreciated. Indeed, our society celebrates their transfer and absorption into automatic devices and smart technologies. The resulting existential remoteness from the world is all the easier to accept because, by common consent, the world is more complex, uncertain and dangerous than it used to be. Faced with an unpredictable environment, the digital recoupment of distance through new sense-making tools and smart technologies appears, if not providential, then at least fortuitous. Penetrating all areas of personal, national and international life, this transfer has been rapid, complete, and affects us all. Yet – making it all the more remarkable – it is only now, when all but complete, that this enfoldment has started to attract the wider critical attention it deserves (Morozov 2013; Carr 2015; Foer 2017; Taplin 2017).

In trying to render some of this historic capture accessible, this book has focused on the changing global North–South interface. In particular, it has interpreted humanitarian disaster as part of a boomerang effect, or feedback loop, interconnecting the North and South during the transition from Fordism to capitalism's new network economy. In relation to the computational turn, and the transfer of human agency to self-acting technologies, humanitarian aid is revealed as a process of anticipation and rupture. To draw this process out, previous chapters have examined the driving spirit of the new economy, delineated some of the transitional forms between knowledge and data, discerned patterns of technological incorporation of precarity, and described the new sense-making tools, disaggregated biopolitics and governmental logics that are now shaping a post-social world.

By way of concluding, the prospect of social automation, which these developments are variously calling forth, is a worthy topic. Central to such automation is the cognitive turn – that is, the privileging of the unconscious and automatic thought processes upon which *Homo inscius* is reliant. Building on the preceding chapters, the Conclusion outlines current attempts to streamline the social reproduction of precarity through attentive cognitive designs and smart feedback. Of particular importance, however, is how these initiatives – together, the general thematic of the book – dovetail with the long postmodern trope of the caretaker society. Having solved all major social and political problems, all that remains for such a society is the continual round of piecemeal technical adjustment. In such a world, the allure of design has vanquished politics and dreams of radical change. There is, however, a paradox at the heart of commercial connectivity. The remoteness and consequent complacency of elites is deepening at the same time as unprecedented societal polarization, fragmentation and visceral anger spread.

Cognitive Turn

The World Bank's 2015 World Development Report is entitled *Mind, Society and Behaviour*. Tracing a link to the work of Fredrick Hayek (World Bank 2015: 5), the Report is a major statement on the importance of cognitive science to international development (Alcock 2016). *Mind, Society and Behaviour* has been welcomed by developmentalists keen to draw parallels between the Bank's work and current leading-edge initiatives among NGOs, such as *Doing Development Differently*[1] and *Thinking and Working Politically*[2] (Green 2014; Ramalingam 2014). What can be called cognitive development, aspects of which were anticipated during the fantastic invasion (chapter 5), has moved to the foreground, in parallel with the emergence of crisis informatics and humanitarian design (chapters 10 and 11). Indeed, cognitive science interleaves these initiatives. It provides a neurological framework that is receptive to and actionable by behavioural technologies. When addressing precarity from a cognitive perspective, the terms 'developmental' and 'humanitarian' lose their former distinctiveness. Since the chronically poor and the disaster-affected are one and the same people and constantly change place, these terms have blurred and become interchangeable. Using the Bank's Report as a point of departure, the Conclusion first examines the contribution of cognitive science to optimizing the social reproduction of the precariat under conditions of permanent emergency.

Mental precarity

Mind, Society and Behaviour brings two key elements to international development. First, as seems common following the scare of the 2008 financial crisis, is a rejoinder that, despite everything, global capitalism does not require any fundamental change or reform. Claims that poverty results from the political and economic system being stacked against the poor, which thus could be solved by 'quotas or a large-scale redistribution of resources', are rejected as incomplete (World Bank 2015: 80). It is argued that redistribution would not 'address the cognitive resources required to make a decision, especially when material resources are in short supply and when people's willingness to act upon their desires may be constrained' (2015: 80). While one could be forgiven for thinking that the purpose of redistribution would be to provide such resources and easements, this strange argument is only tenable if you believe that the poor, compared to the rich, are cognitively challenged. But, there again, that's the point of the Bank's Report.

Following from this, the second important element is the redefinition of poverty in terms of 'bandwidth', or the sum of the enfolding infrastructural or environmental mental aids and resources – or lack of them – that are available to a sentient being. Mind and behaviour are thus determined by one's social milieu. An example is given of a poor indebted farmer, the harvest still months away, pressed to decide whether to make a long-term investment in the education of a child. This is happening when there is a hole in the roof, the kerosene has run out and finding clean water is a constant effort. In addition, his neighbour is expecting help with pressing medical bills because the farmer's family received similar support from his neighbour in the past.

From this behavioural perspective, poverty is not a structural or social outcome: it's a personal experience. It is the result of the constant grind of having to make hard choices: educate a child, fix a roof or invest in communal reciprocity? Relentless hard choices 'in effect tax an individual's bandwidth, *or mental resources*. This cognitive tax, in turn, can lead to economic decisions that perpetuate poverty' (emphasis added, 2015: 81). At the same time, off-grid environments that lack regular water, electricity or sanitation services also increase the cognitive levy. A high mental tax creates poor frames of thought and makes for impaired decision-making. For the World Bank, thinking is a zero-sum game. The more 'bandwidth' the poor consume in their daily grind, the less they have for making important decisions. Presumably, the greater the privation, the more mindless the poor become. Reducing this cognitive tax – and cutting tax is always a popular neoliberal move – consequently leaves more bandwidth for better decision-making. And better decision-making on the part of the poor, as mentioned above, fortuitously negates the need for 'a large-scale redistribution of resources'.

While it is tempting to reject this ideological edifice as so much intellectual detritus, since *Mind, Society and Behaviour* has been warmly received among high-bandwidth elites, the reframing of poverty as a cognitive tax has to be taken seriously. Aside from reconfirming the centrality of neoliberalism's *Homo inscius*, it sidelines understanding poverty as structurally implicated in the production of wealth in favour of seeing it as a personal experience. Behavioural economics thus draws freely on the sociological and anthropological literature to emphasize the constant hassle and uncertainty of precarity. Reflecting the post-humanist turn, it highlights the direct and unmediated relationship of the poor to the perturbations of their enfolding environments. In one respect, this is useful. Behavioural economics does, indeed, draw attention to the constant daily struggles, hardships and insults endured by an expanding global precariat. *Mind, Society and Behaviour* is full of such examples, like the poor indebted farmer previously cited. This evidence, however, is not being used to call for significant social change or radical reform. In fact, the opposite takes place. During the nineteenth century, a growing professional awareness of the plight of the poor helped to catalyse a long process of political mobilization and incremental reform, eventually resulting in universal suffrage and the welfare state (Rabinow 1995). At a time when global inequality is at record levels and new forms of post-social servitude and abjection are appearing (LeBaron & Ayers 2013), the reactionary repositioning of poverty as an experience open to cognitive massage now uses a narrow behavioural empiricism (see chapter 11) to reject a structural dimension to want, and thus any need for meaningful change.

Feedback

Cognitive development and crisis informatics interconnect within the thematic of early warning. Following the rapid spread of mobile telephony, the discovery of disasters as distributed information systems builds on and radically extends the analogue model of early warning that first emerged in the 1970s and 1980s (see chapter 5). Disasters generate abnormal behaviour patterns that, as distinct signals and alerts, can be recorded, analysed and acted upon. The daily hassles faced by a connected precariat as they make difficult choices likewise leave a data trail. Living on the edge between economy and disaster, the informatics of precarity usefully sums up the UN's Global Pulse project. Made possible by remote sensing and mobile telephony, working in near real-time, you can now 'figuratively take the pulse of communities' (UNGP 2009: 8): 'disaster affected communities have become increasingly "digital" as a result of the information revolution. These new digital technologies ... are evolving a new nervous system for our planet, taking the pulse of our social, economic and political networks in real-time' (Meier 2013b: 3).

Big Data analytics and, importantly, the feedback loops created by mobile telephony and interactive devices have built on the basic early warning model and, as the trajectory of Facebook suggests (Foer 2017), transformed it into a powerful tool of behavioural analysis, prediction and manipulation. The notion of humanitarian disaster has expanded beyond its traditional focus on major political and environmental scourges and upheavals. As discussed in chapter 9, smart technology levels downwards, encompassing the micro-level world of hard choices among the precariat as it enfolds the everyday workings of a connected world. If living on the edge, as reflected in the hard choices made by the precariat, can be recorded and visualized as behavioural patterns, behavioural economics seeks to reverse-engineer this situation through feedback.

Cognitive development involves the packaging and delivery of designed, context- specific information to optimize precariat decision-making. A feedback loop has four distinct stages. First, behaviour must be captured, stored and algorithmically analysed. Second, the returned information must be personalized to the individual or group in a way that resonates emotionally. Third, this value-added information must illuminate a way forward. And, finally, there must be 'a clear moment when the individual can recalibrate a behaviour, make a choice and act. Then that action is measured, and the feedback loop can run once more, every action stimulating new behaviours that inch us closer to our goals' (Geotz 2011).

Behavioural economics is premised upon the constant tailoring and readjustment of information 'to fit the human body and its cognitive abilities' (World Bank 2015: 2). Not only does the feedback of value-added information free individuals from the burden of having to sift through gigabytes of noise and distraction, as discussed further below, but also the need to make decisions can be timed to occur when the poor are at their most attentive. This approach informs, for example, the idea of *Doing Development Differently* through processes of adaptive design, an initiative captured in the slogan 'from best practice to best fit' (Ramalingam 2014).

Streamlining has already been discussed in relation to digital infrastructure – in particular, how smart technology levels downwards by shaping itself to the inequalities and differences encountered. The external tailoring of the infra-informational environment of the cognitively challenged subject to shape desired behaviour goes to the heart of algorithmic governmentality (Rouvroy 2012). The authority for such affect management rests upon a practitioner consensus regarding the 'necessary ignorance' of neoliberal's *Homo inscius*. For behavioural economics, human consciousness divides into 'reflective' and 'automatic' systems. The former relates to the use of reasoned deliberation to achieve conscious goals. The latter, however, is held to be far more important in the shaping of actual behaviour. Operating below the level of conscious reason or recognition, it hinges on the automatic play of unconscious heuristics, environmental cues, mental shortcuts or the unreasoned operation of group preferences and shared models (World Bank 2015: 3–4).

While having serious implications for transparency and democracy, such assumptions and related technologies remain politically unchallenged (Alcock 2016: 102–10). Who decides what optimal behaviour is? From predictive marketing and 'nudge' politics through to enfolding parametric architectures (Thaler & Sunstein 2008; Spencer 2016a), these technologies seek to shape our cognitive environment. Amid accusations of fake news and social media manipulation, the election of Donald Trump suggests that these technologies have also disappeared into the wild, so to speak (Cadwalladr 2017). The application of behavioural informatics, honed in the personalized consumer markets of platform capitalism (Srnicek 2016), to precarity is thus not unique to the interface between economy and disaster. It is more a question of market expansion as third-wave electronic globalization now folds itself into the sub-prime tele-economic conditions of the global South. What we are seeing is a double movement. On one level, this is an extension of cognitive capitalism (Boutang 2011 [2008]) into the relations and interactions that maintain the social reproduction of the global South's precariat. At the same time, given the post-social context of precarity in the South and its vast extent, there is also the experimental level of the boomerang effect whereby new forms of social automation and nomadic servility can be anticipated.

Optimizing reproduction

In chapter 8, it was argued that the current wave of automation in the global North is reducing the number of professional and middle-class jobs that depend on logical or algebraic modes of thought. Machine learning finds such tasks relatively easy. More difficult to master are lower-level sensorimotor skills that rely on perception, mobility and dexterity (Joshi 2017). Low-waged, insecure, temporary and often technologically stagnant 'service sector' jobs reflecting such skill sets is one of the few areas of employment that is expanding. Rather than a temporary phenomenon, it is contended here that regional variations on such activity, if available, reflect the future of work for the global majority. In this respect, Luc Boltanski and Arnaud Esquerre's (2016) challenging argument concerning the emergence of 'enrichment economies' in the 'post-industrial' global North, which are characterized by 'a patrimonial class of growing importance on the one hand, and a badly paid, insecure precariat on the other', deserves serious attention (2016: 36, n7). Reflecting the irony of the negative dialectic, as capitalism adjusts to make the most of its precarious outlook, thousands of platform entrepreneurs, politicians, software engineers, financial innovators, academics, development practitioners and humanitarian designers, for the best of intentions, are rising to the challenge of making an entropic and barbaric future liveable.

CONCLUSION: AUTOMATING PRECARITY

For the World Bank, those aspects of social reproduction among the precariat that can be cognitively targeted for behavioural optimization include household savings, energy consumption, educational priorities, mental and physical productivity and, not least, maternal and child health (World Bank 2015: 2). These elements span the individuated post-social biopolitics that reproduces the cheap, territorially immobile and dispensable low-level sensorimotor skills that drive the vast informal economies of the global South. *Mind, Society and Behaviour* emphasizes in several places that cognitive techniques are attractive because they cost relatively little, need not be complex and are already widely practised in the private sector. With regard to existing aid programming, rather than requiring radical change, it is largely a question of 'nuances of design and implementation' (2015: 3). Taking 'the cognitive taxes of poverty into account' (2015: 81) might simply involve changing the timing of cash transfers, altering the labelling on foodstuffs, simplifying processes or service take-up, sending out regular reminders, marketing new social norms or 'reducing salience of stigmatised identities' (2015: 3). The cognitive tax on the precariat could be significantly reduced by shifting the timing of critical decision-making regarding, for example, education, health or employment 'away from periods when cognitive capacity and energy (*bandwidth*) are predictably low', and, alternatively, by 'targeting assistance to decisions that may require a lot of bandwidth' (2015: 81).

Another important dimension for easing the burden of thinking relates to infrastructure. Having to daily exert a great deal of mental energy just to access such basic necessities as food and clean water means the precariat 'are left with less energy for careful deliberation than those who, simply by virtue of living in an area with good infrastructure and good institutions, can instead focus on investing in a business or going to school committee meetings' (2015: 13). Thus, the absence of a universal fixed-grid 'like piped water, organised child care, and direct deposit and debit [accounts for] earnings – encumbers those living in low income settings with a number of day-to-day decisions that deplete mental resources even further' (2015: 81). For *Mind, Society and Behaviour*, cognition-aware policy instruments 'such as cash transfers and the development of infrastructure, institutions, and markets' could serve 'to lessen the distractions and cognitive burdens of poverty' (2015: 81). In this respect, cognitive development provides a neurological rationale for the personalized and attentive humanitarian objects discussed in chapter 11 that have been specifically designed to support wild forms of nomadic off-grid survival.

Just as the objectivity of the satellite image relieves the public of the burden of having to read human rights reports, the World Bank similarly sees behavioural economics as easing or streamlining the task of thinking among the poor. Everywhere, it would seem, mnemonic technologies and attentive practitioners are striving to free us from the efforts, burdens and risks of thinking. From disasters to the gig economy, feedback loops are optimizing us logistically to be in the right place at the right time (Reid 2006). We are now all enrolled in the politically unremarked – and, for the most part, unrecognized – creation of what Bernard Stiegler has called *automatic society* (Stiegler 2016). Regarding the social reproduction of the precariat, we can see anticipatory forms of cognitive streamlining, value-added feedback and aid automation in the global South that are moving in this direction. In biopolitical terms, informal economies are being envisioned and celebrated as capable of self-reproduction and self-organization under conditions of permanent emergency. Moreover, apart from some design initiatives, this requires minimal effort or outlay from the global North and, especially, no need for radical change. In this respect, the ontologies and technologies discussed in this book reproduce the long-anticipated self-adjusting society. This time, however, rather than this being a subject of imagination, we are now seeing early attempts to operationalize it and make it real.

Caretaker Society

At the end of the Cold War, the West's economic and political ascendancy seemed assured. Reflecting this spirit, in 1989 Francis Fukuyama published his polemical article 'The end of history?' (Fukuyama 2002). For Fukuyama, this ending was signalled by the demise of grand theories and the growing suspicion and rejection of struggles that continued to call for recognition or demand justice – especially those willing to use violence to press such demands. Although the coming world would be more peaceful and secure, lacking such history-making sacrifice, it would not be a particularly exciting place. A self-satisfied inertia would descend as societal administration becomes routinized around continuous, 'economic calculation, the endless solving of technical problems, environmental concerns, and the satisfaction of sophisticated consumer demands. In the post-historical period there will be neither art nor philosophy, just the perpetual caretaking of the museum of human history' (2002: 178).

At the end of history, the destructive clash of divergent interests and competing world-views is replaced by progressive moderation and political pragmatism (see also Sloterdijk 2013 [2005]). This imaginary, however, is not new. In 1960, Daniel Bell published his celebrated book *The End of Ideology: On the Exhaustion of Political Ideas in the Fifties* (Bell 2000 [1960]). Having failed to stop World War II, the grand humanist theories of the nineteenth and early twentieth centuries were argued to have outlived their usefulness. While ideology still characterized the often-violent political emergence of the Third World, more pragmatic and restrained beliefs were taking root in the West. Among sensible people, reason had transmuted into reasonableness. Rather than radical change, piecemeal technical adjustments would now shape future society.

The thematic similarity between the 'end of ideology' and the 'end of history' is striking. In addition, both emerged at a time when the West temporarily stood victorious after major global struggles against external political enemies. It was assumed, moreover, that these successful struggles had also helped to resolve the important internal social and political issues of the time. In both cases, the future was one of piecemeal technological adjustment and, to use Fukuyama's phrase, perpetual caretaking. Without the clash of ideologies or recalcitrance of history, society is seen as sliding into a culture of complacency. While both propositions were derided at the time, the thread of smugness that connects them suggests an enduring celebratory role in excess of any explanatory power they may have. Appearing a half-century after Bell's book, and almost two decades since Fukuyama's contribution, Bruno Latour (Latour 2008) revisited the caretaker trope in a conference paper entitled 'A Cautious Prometheus? A Few Steps Toward a Philosophy of Design'. Having previously pronounced the death of grand narratives and critique (Latour 2004), the conference paper rehearses how the design principle has supplanted a revolutionary political urge. One is tempted to suggest that, in the figure of a *cautious* Prometheus, a self-satisfied world trapped in a spiral of technological lassitude has now found a suitable anti-hero.

As an elaboration of his critical stance towards modernity (Latour 1992), Latour speculates on how the separation between materiality and design that characterized modernity has weakened and blurred. Since encountering the scale and complexity of the ecological crisis, we are all now designers rather than modernizers. Across a wide arc of operational discourse, designer attitudes reflecting such sentiments as 'attachment, precaution, entanglement, dependence and care' have all but replaced earlier and more reckless commitments to 'emancipation, detachment, modernization, progress and mastery' (Latour 2008: 2). In this fashion, Latour asserts that the ontopolitics of the design principle – the need to work with the world as it is, rather than how it ought to be – has now effectively supplanted the idea of revolution. It would be difficult, he suggests, to find feelings or sentiments like humility, attentiveness or ethics as having played any formative role in the revolutionary movements of the past. The caretaking of design sits awkwardly with a Promethean urge to raise and construct – and, in consequence, to tear asunder and destroy.

Latour is no doubt correct to assert the centrality of design within the postmodern canon. In that case, alongside the end of ideology and history, since they are the basic ingredients of political life, the caretaker society now adds the death of politics as well – at least, that is, a politics that is staked on the commons between reality and the world, and dares to use the former to critique the latter.

There are differences, however, within this enduring trope. Ideology and history ended at moments of geopolitical victory over the external enemies of fascism and totalitarianism. The end of politics seems to mark a different kind of reckoning: less a victory, more a defeat. In particular, it signals that failing of nature by society that has produced humanity's negligent own-goal of the Anthropocene (Bonneuil & Fressoz 2016) – the discovery of which, incidentally, has been the most important to date for predictive computer modelling (Edwards 2010). The current embrace of caretaking and design does not result from a triumph over external enemies. It is more a retreat from radical change because, in the last analysis, the Promethean 'victories' of World War II and the Cold War were in fact Pyrrhic. In the drive for industrial mastery, did they not call forth the ecological crisis? An ontopolitics of caretaking design is now necessary because, as history shows, left to themselves, humans make things worse. Whereas past victories were geopolitical and ideological, this failure is more behavioural and biopolitical. Moreover, this time around, caretaking benefits from the fortuitous arrival of Big Data and new automatic sense-making tools and logistical platforms that, because humans are now out of the loop, are hyper-objective (Spencer 2016b).

The enemy is now internal; it is human behaviour itself (Reid 2006). The tragic ignorance of the War on Terror lies in the choosing to outlaw a set of behavioural patterns (Chamayou 2015 [2013]). Given the impossibility of victory over such an enemy (or, perhaps worse, that one is being enforced regardless), one cannot help feeling that today's caretaker is more disquieting than its earlier rather nerdish iterations. The urge to record, monitor and adjust to ensure things remain within accepted parameters feels more extreme and determined. As discussed in chapters 8 and 9, in the techno-pastoral, this attentive caretaker has even created its own aesthetic of a timeless precarity. However, as the transfer of reason and human agency to self-acting technologies increases, the homely entropy of this aesthetic loses its gloss. Rather than empowered aid beneficiaries resiliently adapting to endless emergency as they smile from agency advertising brochures, the post-social world now in formation has more the appearance of expansive ruined and wild landscapes riven by desperate struggles against new and emerging patterns of off-grid servitude.

Paradox of Connectivity

When technoscience replaces politics as a means of international problem-solving, there are repercussions. As argued in this book, there is a formative ontological and epistemological relationship between a world seen as complex and dangerous and the data-based sense-making tools used to establish and understand this condition. For conventional wisdom, however, they appear unconnected in any intrinsic or formative sense. The utility of machine-thinking, for example, lies in its ability to uncover objectively pre-existing complex relations in the outside world that are otherwise beyond human comprehension. At the same time – and this is the rub – since those complex relations pre-exist computer analysis, the same machine-tools can be confidently used to resolve the dangers only they can disclose. For this computational capture of responsibility to work, it helps if one believes that new technologies are in fact 'new' – that is, they have somehow leapt immaculate and ready-made into the present. As analysed in relation to the boomerang effect, however, these technologies have a long history of appropriation. They are implicated not only in contested ontological and methodological changes regarding how the world is understood, but also in a profound shift in the nature of capitalism. In many respects, the computational turn feels more like an arrival – or, better, an epistemological closure – rather than a new beginning. The corporate machines, as it were, are now consolidating the two or three decades of anticipatory cybernetic and behavioural ground work that preceded their arrival. As a way of gaining greater perspective, the idea of a *paradox of connectivity* is useful.

The paradox draws on the contrast, used to introduce this book, between the revolutionary optimism of the past and the political pessimism of today. It takes into account not only the pulling apart of previously culturally mixed societies but also the resulting loss of familiarity. The striation of international space into fast and slow lanes is important, as is the blurring of economy and disaster to produce a global precariat – perhaps the new economy's single greatest achievement. Precarity draws attention to the blurring of North–South dynamics, growing inequality between and within countries, declining living standards, jobless growth, the casualization of work and the migration crisis. Young people everywhere no longer expect the life-chances their parents, or even grandparents, enjoyed. However, rather than from the Munich Security Report (MSR 2017) cited at the beginning of this book, the paradox draws more from the key parameters of Pankaj Mishra's (2017a) *Age of Anger*. Instead of a rising tide of illiberalism, authoritarianism and populism *per se*, these are the epiphenomena of a more fundamental change: a political rejection of the bankruptcy of progressive neoliberalism and the cosmopolitan values and universalism it espouses. A new global society is struggling to free itself from the old, with all the problems of disjunction, incoherence and contradiction that this necessarily involves.

While 'isolation' can be said to be the opposite of 'connection', they have no separate or mutually exclusive existence. They always exist together and are socially constitutive of each other (Read 2016). How a person connects – which in terms of the technology involved is historically given – defines the quality of their isolation or remoteness. Indeed, following Arendt's argument regarding world alienation, we

CONCLUSION: AUTOMATING PRECARITY

can suggest that remoteness and connectivity are directly related; the greater the connectivity, the more distance or isolation. The paradox of connectivity, however, introduces a real-world consideration. It holds that the greater the reach and speed of connectivity, the more ground friction is generated. How speed and distance translate into ground friction lies in the history of connectivity – especially, its role in the trajectory of contemporary capitalism. Connectivity generates ground friction because, as argued in chapter 2, it is integral to new forms of network exploitation and the abjection of the slow and immobile.

Technoscience displaces or side-steps difficult political problems, such as the need for a new formula for sharing the world with others, by transforming them into easier or more do-able technical challenges – for example, the need to make sense of distant disaster zones now deemed unsafe for international aid workers; or how to provide access to clean water without a fixed infrastructure; or ways to optimize the decision-making capacities of the poor while avoiding any genuine democratization or redistribution of resources. This process of displacement marshals the positive energy and empathy of innumerable researchers, entrepreneurs and politicians in the quest for solutions. However, the hard political problems remain. Moreover, since these problems easily become compacted and amplified, the main result is to produce new and recurring rounds of ever-distant global challenges. One could say that the negative dialectic, or the sequential journey of capitalism towards an entropic barbarism, proceeds on the basis of good intentions and the best evidence available.

Since the 1980s, as connectivity has increased, the number of physical and legal barriers preventing the free movement of people have multiplied, social and economic inequality has grown and the world of work has been casualized. At the same time, recalcitrance, anger, political push-back and international no-go areas have spread. The paradox of connectivity goes against the grain of conventional wisdom. As stakeholders in the new forms of appropriation and governance that it has made possible, for the status quo increased connectivity is the best tool we have for solving global problems. While this may yet still be the case, within the present framework of the commercial ownership and control of the data, algorithms and smart technologies involved, such an outcome is unlikely. The paradox of connectivity lies at the heart of a design-dependent caretaker society. Connectivity is a pharmacon. It is simultaneously a benefit and a scourge (Stiegler 2016). It is useful for a post-social capitalism but a disaster for the global precariat in formation. As the remoteness and complacency of elites deepens – and the World Bank's view of cognitive precarity is a case in point – continuing polarization, fragmentation and anger appear set to continue shaping the future.

NOTES

Chapter 1 Introduction – Questioning Connectivity

1 For a critical review, see Mishra (2017a).
2 The terms 'global North' and 'global South' are used here figuratively. Loosely associated with earlier modernist distinctions between developed and underdeveloped countries, they no longer imply any fixed geographical or social homogeneity. Their use, however, serves to retain the sense of a historic political and economic division that continues to produce global power and distribute life-chances unequally.
3 'Ground friction' is a generic term for anything that acts to slow the ease or speed of terrestrial movement and circulation. This can range from bureaucratic hurdles, insurance requirements and restrictive regulations through to political resistance, insecurity and weak public institutions.
4 This is a notion borrowed from Paul Virilio.
5 I am not seeking to detract from modernism's colonial violence and periods of murderous excess. The focus here is that reappropriation of the modernist legacy that, itself, has produced new forms of control and domination.
6 Stephen Graham has also used the notion of the boomerang effect in relation to the development and perfection of Western security techniques in the global South (Graham 2013).

Chapter 2 Against Hierarchy

1 In the development of contemporary global capitalism, the first wave of globalization took place between 1870 and 1914 (Hirst & Thomspon 1996). This initial step-change in space–time compression was made possible by developments in steam and communication technologies. In the days of sail, the journey from England to India via the Cape of Good Hope, for example, could take up to 6 months. Replies to letters could take a year or more to arrive. Together with the appearance of the steam ship, a major

globalizing effect occurred with the opening of the Suez Canal in 1869. From months, the passage to India became an increasingly predictable journey of 2 or 3 weeks. Regarding telecommunications, between the 1850s and 1860s undersea cables first connected Britain with continental Europe before crossing the Atlantic and then moving down the Red Sea to India and beyond. By the mid-1870s, a telegram sent from India could be expected to arrive in London some 3 hours later (Headrick 2012: 22).

2 Other names include, for example, 'cognitive capitalism' (Boutang 2011) or 'platform capitalism' (Srnicek 2016). These concepts are indicative of important characteristics of the present.

3 For the lower classes, capitalism promised a spatial liberation from constraining ties to the land, village and family, a promise realized in the rapid growth of towns and mass emigration from Europe to the New World and the colonies. During the nineteenth century, 60 million left in this outward dynamic, which today, as the barriers to migration attest, has reversed.

4 Contrary to the claimed 'embourgeoisement' of the working class that formed a central part of the May '68 critique, the middle classes benefitted most from the post-World War II social-democratic settlement and the rapid expansion of professional and managerial employment associated with welfare-Fordism.

5 Except for limited pattern recognition, humans are incapable of understanding society due to its alleged complexity. Fortuitously, however, the market compensates for human ignorance. The price mechanism functions like a computer and is able to achieve optimal resource allocation through its powers of spontaneous self-organization.

6 Crawford Holling's celebrated article on ecological resilience is a critique of equilibrium theory based upon research on predator/prey relations. See Holling (1973).

7 For an updating of 'proletarianization' in the digital age, see Stiegler (2010).

CHAPTER 3 ENTROPIC BARBARISM

1 Some elements of the New Left would come to embrace Maoism, and, before the situation was better understood, even the Khmer Rouge.

2 Following a decade of growing numbers of military advisers, US ground forces eventually became directly operational in Vietnam in 1965.

3 In the UK in the mid-1960s, the Labour Party was instrumental in introducing and giving intellectual coherence to the control of immigrants from Britain's colonies and former colonies (Duffield 1984). In practice, this restriction effectively severed any legacy of solidarity that existed between the labour movement and the forces of Third World revolution. In place of solidarity, a compensatory regime of external development assistance was instigated at the same time as deploying an internal one in the form of a domestic 'race relations' industry (Labour Party 1964; see also Duffield 1988).

4 At the tail-end of this trend, my first academic job was as Research Fellow

in the Department of Folklore at the University of Khartoum between 1976 and 1979.
5 For a critical examination of continental philosophy from the perspective of the negative dialectic, see Noys (2012).
6 Appearing in 1916, two years before the Russian Revolution, the idea that the choice between 'socialism or barbarism' had replaced the nineteenth-century alternative of 'capitalism or socialism' is attributed to Rosa Luxemburg (Luxemburg 1916). For Luxemburg, the cataclysm of World War I, together with the mass conscription of workers and political capitulation of International Social Democracy to the militarism of the imperial powers, was a demonstration that bourgeois society had undergone a 'regression into barbarism'. The choice was either the triumph of imperialism and spread of total war or, alternatively, a victory for socialism. In Europe, this dilemma would be resolved in the negative with the spread of totalitarianism – an outcome that cost Luxemburg her life.
7 Concerns relating to entropy and to society moving towards a state of indistinction find an interesting resonance in the work of the philosopher Giorgio Agamben. He has drawn attention to how the declaration of successive states of emergency within recent European history has culminated in a condition of permanent emergency (Agamben 2005). Moreover, this condition has led to an increasing political 'indistinction' between the formative registers of liberalism (Agamben 2013). A connection between Agamben's work and the negative dialectic, especially highlighting the emergence of a new mass form of 'neo-fascism', is suggested here.
8 By 1980, second-order cybernetics had morphed into an epistemology that sees the world as a set of informationally closed systems. Organisms respond to their environment in ways determined by their internal self-organization. Their one goal is continually to reproduce the organization that defines them as systems (Hayles 1999: 10–11). By this time, cybernetics was on course for its contemporary enabling relationship with neoliberalism and, eventually, resilience-thinking and the pure factuality of post-humanism.

CHAPTER 4 BEING THERE

1 As a measure of the academic freedom existing at the time, the publication of Herbert Marcuse's *One Dimensional Man* is instructive. The 1968 paperback edition had as its cover endorsement, 'THE MOST SUBVERSIVE BOOK PUBLISHED IN THE UNITED STATES THIS CENTURY' (original emphasis). Marcuse pulls no punches in condemning capitalism and calling for revolutionary change. Significantly, however, he acknowledges and thanks several mainstream foundations, including the American Council of Learned Societies, the Rockefeller Foundation and the Social Science Research Council, for their grant support. The first edition of *One Dimensional Man* sold 300,000 copies and encouraged and helped articulate innumerable acts of academic protest and campus resistance. It is impossible to imagine such autonomy in today's much more controlled, managed and indexed academic environment. The two situations are, quite literally, worlds apart.

2 This dismissal has tended to occlude the history of the unincorporated. The mainly university-based May '68 movement coincided with growing industrial unrest in Europe. By the mid-1970s, with immigrant workers playing an important role, an epidemic of unofficial 'wild cat' strikes threatened to wrest shop-floor control from corporate managers (Duffield 1988). The several decades of rising wages that ordinary workers enjoyed under Fordism drew on such unincorporated shop-floor struggle. Following the deindustrialization of the 1980s, the unincorporated have suffered a steep decline in living standards, with many joining the ranks of the left-behind.
3 Between 1968 and 1971, I was a politically active undergraduate in the Department of Sociology at the University of Sheffield.
4 Aspects of this applied anthropology re-emerged in Iraq and Afghanistan in the mid-2000s with the US military adopting a cultural-functionalist understanding of its adversaries (McFate 2004). Drawing criticism from the American Anthropological Association (AAA), social scientists were recruited to map the 'human terrain' as part of US counterinsurgency measures in Afghanistan (Gonzalez 2008).
5 If you set off at first light you could hope to be in Khartoum by early evening as darkness fell. Following the extension of the road from Wad Medani in the late 1970s, the journey time has halved to around five hours. Moreover, it can now be completed in an air-conditioned bus.
6 *Fellata* is a longstanding Sudanese generic term, usually having derogatory connotations, for settlers of West African descent.
7 Sultan Abu Bakr Mohammed al Tahir died in 2016. Abu Bakr was a direct descendant of the Mohammed Bello Mai Wurno (d.1944), who withdrew to Sudan after the resistance of the Sokoto Caliphate was crushed in northern Nigeria by the British at the 1903 battle of Burmi. Until the abolition of native administration in 1971, the Maiurno Sultanate had exercised some administrative authority over Fulani settlements between the Blue Nile and Dinder rivers, and the Umborro nomads further south. Since independence, however, the Sultanate had been in decline. During the 1990s, some administrative powers were restored. When I visited Maiurno in 2014, the Sultan's guest rooms had been refurbished. The dilapidation and leaking roofs of the 1970s had gone.
8 As my academic life took a different course, this material lay forgotten in various drawers and attics for many years. In January 2014, twenty cassette tapes of recordings were given to the Institute of African and Asian Studies at the University of Khartoum for digitalizing and cataloguing.

Chapter 5 Fantastic Invasion

1 I am grateful to Dan for pointing out that the original term is from Joseph Conrad's *Heart of Darkness*: 'And outside, the silent wilderness surrounding this cleared speck on the earth struck me as something great and invincible, like evil or truth, waiting patiently for the passing away of this fantastic invasion' (Conrad 2002: 125).
2 Mozambique, for example, had its own NGO invasion during the war years

of the 1980s. From around 70 in 1985, they had increased to 180 by 1990 (Duffield 2007: 87).
3 For an early 1990s literary expression of this interventionary cosmopolitanism, see Postlewait et al. (2006).
4 Boltanski and Chiapello (2005: 350) point out that the displacement of political activism into humanitarian action was, in some respects, surprising. The May '68 movement had generally attacked the 'charity' of aid agencies as hypocritical, complacent and displacing the need for political change. This position can be seen, for example, in George (1976) and Dumont and Cohen (1980); for a contemporary version, see Furedi (1994).
5 By the mid-1980s, Oxfam had been around for forty years. Its first Africa grant was made in 1953. Until the mid-1960s, it mainly worked with the colonial administrations in southern Africa (Black 1992). In terms of my recruitment, the area expertise gained through ethnographic fieldwork in Sudan was then still valued.
6 See https://en.wikipedia.org/wiki/Telex.
7 NGOs had pigeon-holes in the reception where in-coming telex traffic was left for collection. This provided an opportunity for aid agencies to skim each other's communications. Convenience and security improved in 1986 when Oxfam got its own office Telex line. However, communications remained subject to Sudan's decaying terrestrial infrastructure and kick-back culture. This dependency was partly off-set by the now closed Portishead Coastal Radio (see http://en.wikipedia.org/wiki/Portishead_Radio), which allowed overseas terrestrial short-wave transmitters to patch into the UK landline telephone network. Portishead, however, was insecure, as third-parties could eavesdrop the station's open-channel broadcasts. At this time, it was also still possible to monitor embassy and UN local communications with an off-the-shelf VHF scanner, since little traffic was then encrypted.
8 The entrepreneurial political atmosphere of the time is well captured in Nicholas Winer's Ethiopia-based aid thriller, *The Tethered Goat* (Winer 2008).
9 The weakness of direct humanitarian action, as with all forms of autonomy, is a susceptibility to capture by the apparatuses that are threatened (Agamben 2009). Direct action in Sudan threatened NGO HQ authority, the Sudanese state and the prerogatives of the international aid industry. Regarding the first, with the appearance of more effective telecommunications, the relative autonomy of the field was lost. The Islamist revolution in 1989 enabled the state to assert its authority over NGOs (Karim et al. 1996). Capture by the aid industry took the form of increasing donor and UN managerial control over NGOs. Donors tightened up their subcontracting procedures. Inclusion within the UN's Operation Lifeline Sudan (1989–2005) required formal NGO compliance (1996). Autonomy was traded for access to logistics and funding.
10 The speed and ideological completeness of the fantastic invasion was disorientating. While working for Oxfam, I never returned to Maiurno. This was justified in terms of lack of time due to the demands of the job. However, the change in power relations resulting from the fantastic invasion was

troubling. Overnight, as it were, friends had become potential aid beneficiaries. The thought of confronting this dissonance was unsettling and resulted in taking the easy way out. My failure to return, however, was noted, and has only been assuaged in recent years.

11   As part of the same progressive movement, the criticism of the refugee camp was coterminous with a similar dismantling of the old mental asylums in Europe (Cowen 1999).

12   Liberalism and neoliberalism differ at a number of levels. One of these relates to security. While accepting its inevitability, liberalism was historically susceptible to political forms of protection against uncertainty, including insurance and the welfare state (O'Malley 2009). Neoliberalism, on the other hand, not only accepts risk, it works to remove protection in its celebration of risk as essential for life and profit (Cooper 2008).

13   The marked commercial and political dominance of the riverain Arabs would not gain such prominence again until the Darfuri Islamist group Seekers of Truth and Justice (STJ 2000) published its celebrated *Black Book: Imbalance of Power and Wealth in Sudan*.

14   Before joining Oxfam, for a couple of years I had been Project Leader for a region-wide equal opportunities initiative based in Birmingham Social Services.

15   In the early 1980s, while researching the history of Indian workers in the West Midland foundry industry, I was struck by the visible change in the way that the main trade unions communicated with their members. From the 1950s until around the mid-1970s, trade union newsletters and journals contained detailed overviews of industry trends, policy changes and political commentary. They were substantial documents. From the mid-1970s, however, they quickly reduced in size and format, becoming several pages of mainly colour photographs, charts and graphs interspersed with dialogue boxes.

16   Lacking a standard definition, by the early 1990s a complex emergency was usually regarded as a chronic, multi-causal humanitarian crisis involving varying combinations of political, economic, environmental, conflict and peacekeeping factors that required a system-wide response on the part of the aid industry (UN 1994). Apart from providing a blameless diplomatic language useful for negotiating with perpetrators, the main achievement of this definition was to convert disasters into a multi-agency funding template, thus helping the rapid rise in humanitarian spending during the 1990s.

17   This was supplied by the US National Oceanic and Atmospheric Administration's National Satellite Data and Information Service (NOAA/NESDIS). For the historical background of the famine early warning project, see FEWS NET at https://earlywarning.usgs.gov/fews/overview.

18   Despite setbacks, over the following years, periodic attempts would be made to make terrestrial early warning systems work (Walker 2009 [1989]). When I visited Khartoum in 1999 as a consultant evaluating the EU's emergency programme, I found some NGOs attempting to set up a famine early warning system in Kordofan province similar to that of the mid-1980s. None was aware that this had been tried a decade or so before.

Chapter 6 Livelihood Regime

1 Until the 1980s, industrialization had been largely confined to the import substitution industries in Khartoum North, and the railway workshops in Atbara, together with some refining and processing industries associated with the agricultural export economy.
2 In terms of capacity-building, the project replicates one of the main aims of colonial native administration (Duffield & Hewitt 2009).
3 Projects could also be repurposed according to socio-economic context. In Bosnia during the latter half of the 1990s, for example, where capitalist penetration of the household was unnecessary, projects focused predominantly on conflict resolution, refugee return and confidence building between formally opposed ethnic groups (Duffield 1996).
4 In my earlier work, the NGO project-based model of community development was understood differently from the ideas presented here. The analysis was backward-looking. Since the nineteenth century, for example, community self-reliance has been proffered as a liberal solution to the excess labour that capitalism periodically generates (Duffield 2007). While true, this perspective does not capture the forward-looking spirit of the NGO experimentation with self-sufficiency and self-reproduction during the fantastic invasion. These experiments anticipated capitalism's disembedding of responsibility for social protection. Aware or not, during this period, NGOs were piloting post-social forms of survival that are now boomeranging to the North.
5 I count myself among this number.
6 A policy exchange network managed by the Center for International Development at Harvard University. See https://bsc.cid.harvard.edu/doing-development-differently.
7 An international research initiative run by the Development Learning Programme based at the University of Birmingham. See www.dlprog.org/research/thinking-and-working-politically-community-of-practice.php.
8 It would become the Media Lab in 1985.

Chapter 7 Instilling Remoteness

1 In November 2009, working independently of the aid industry, I spent a month in Khartoum, travelling by foot or using local taxis both day and night. In January 2014, my wife and I enjoyed a two-week holiday there visiting old haunts and friends – again, moving around without let or hindrance. Indeed, we spent a memorable evening in Omdurman at the Muwalid celebration of the Prophet's birthday. The relative safety of the city has also been commented upon in Alex de Waal's blog 'Making Sense of Darfur'. See www.ssrc.org/blogs/darfur/category/darfur.
2 This was in connection with research on risk management among aid agencies (Collinson & Duffield 2013).
3 I am grateful for audience feedback and the anecdotal evidence provided regarding the geographic spread of the fortified aid compound from seminars

given in Rovaniemi, London, Warwick, Cambridge, Amsterdam, Leeds, Bristol, Bradford and Coimbra during 2008 and 2009.
4 For a map of these airstrips, see Relief Web, http://reliefweb.int/map/sudan/south-sudan-roads-airfields-17-jul-2007.
5 These programmes were known as 'negotiated access'. They involved a UN lead agency negotiating on behalf of the aid system with the main warring parties to gain humanitarian access to war-affected populations. OLS is a good example.

Chapter 8 Edge of Catastrophe

1 Interestingly, the policy-makers who authored structural adjustment remained blind to the historic transformation they had initiated (Duffield 2001: 143), just as they would later fail to anticipate the financial crisis of 2008–9 and its global effects (UN 2011: 2).
2 In sub-Saharan Africa, for example, the decline has been modest and, with a headcount ratio (or the proportion of the population living below the poverty line) of 41 per cent, Africa has more extreme poverty 'than all other regions [of the world] combined' (World Bank 2016b: 5).

Chapter 9 Connecting Precarity

1 Estimates of the total amount of electricity now needed to run the digital infrastructure that supports cloud computing suggest an equivalent of that required to illuminate the entire planet in the mid-1980s (Mills 2013).
2 See, for example, http://whiteafrican.com/?attachment_id=4999.
3 I am indebted to an anonymous reviewer for helping to clarify and tease out this relationship.
4 See www.rubensalgado.com/solar_portraits_worldwide.

Chapter 10 Post-humanitarianism

1 During the Cold War, there were several attempts to repurpose military systems analysis and computer-based management applications to solve social problems within the United States – notably, the war on poverty in American cities during the 1960s and early 1970s (Light 2003). Apart from changing the nature of welfare bureaucracy, including rendering mechanical informational and accounting skills redundant, systems analysis actually did little to solve urban poverty. Such programmes, however, helped to forge the celebrated military-industrial complex. In the geospatial sector, while commercial companies grew and multiplied, the military remained the drivers of the technology and expertise. Although low-resolution commercial satellite remote sensing systems were launched, a Superpower high-resolution monopoly lasted until the end of the Cold War (Verjee 2005).

2 Geospatial technology embraces the computer-based acquisition, mapping, modelling, analysis and visual presentation of spatially referenced data derived from GPS applications (Verjee 2005).
3 This process was completed in 2000 when the intentional degradation of the military GPS signal intended for public use was abandoned, giving a tenfold increase in commercial accuracy (Bjorgo 2002).
4 The United Nations Operational Applications Programme (UNOSAT). See https://unitar.org/unosat.
5 RESPOND is a research programme of the Indian Space Research Organisation. See www.isro.gov.in/sponsored-research-respond.
6 See https://ec.europa.eu/jrc/en.
7 Acronym for Environmental Monitoring of Refugee Camps using High-Resolution Satellite Images.
8 Building on the earlier launch of Amazon and Google, it saw the creation of such platforms as Facebook (2004), Twitter (2006), Airbnb (2008), TaskRabbit (2008) and Uber (2009).
9 This was also the argument that prompted the anticipatory move from food aid to supporting coping and livelihood strategies during the 1980s.
10 In 2010, for example, the Standby Task Force (SBTF) comprised 700 technical volunteers in 70 countries, divided into 10 teams providing support for humanitarian responders. In the following 3 years, SBTF was mobilized in more than 22 crises, has worked with UNOCHA, USAID, Amnesty International and Oxfam, and provided support for Syrian diaspora and Sudanese civil society groups. See http://blog.standbytaskforce.com.

Chapter 11 Living Wild

1 See www.bettershelter.org.
2 See https://solar-aid.org.
3 See https://phys.org/news/2010-03-peepoo-bag-sanitary-human-disposal.html.
4 See www.nutriset.fr/en/product-range/severe-acute-malnutrition/plumpy-nut-ready-to-use-therapeutic-food-rutf.html.
5 For Facebook's Free Basics web page, see https://info.internet.org/en.
6 See www.theregister.co.uk/2017/05/23/chinese_etailer_drones_beat_amazon.
7 For example, see e-flux architecture's 'Superhumanity' series at www.e-flux.com/architecture/superhumanity.
8 See https://www.theguardian.com/books/2017/nov/02/fake-news-is-very-real-word-of-the-year-for-2017.

Chapter 12 Conclusion: Automating Precarity

1 See chapter 6, n6.
2 See chapter 6, n7.

BIBLIOGRAPHY

Abu Manga, Al-Amin. 2009. 'The rise and decline of lorry driving in the Fellata migrant community of the Blue Nile'. In *Changing Identifications and Alliances in North-East Africa*, ed. Gunther Schlee and Elizageth E Watson, 139–56. New York & Oxford: Berghahn Books.
Achilles Initiative. 2013. 'Integration training'. Achilles Initiative, www.integrationtraining.co.uk/achilles-initiative.
Ackerman, Robert K. 2001, December. 'Commercial imagery aids Afghanistan operations'. SIGNALonline, www.afcea.org/signal/articles/templates/SIGNAL_Article_Template.asp?articleid=298&zoneid=84.
Ackerman, Spencer. 2015, 16 March. 'ACLU files new lawsuit over Obama administration drone "kill list"'. *Guardian*, https://www.theguardian.com/world/2015/mar/16/aclu-files-new-lawsuit-over-obama-administration-drone-kill-list.
Agamben, Giorgio. 1998. *Homo Sacer: Sovereign Power and Bare Life*. Stanford University Press.
Agamben, Giorgio. 2005. *State of Exception*. Chicago and London: University of Chicago Press.
Agamben, Giorgio. 2008. 'No to biopolitical tattooing'. *Communication and Critical/Cultural Studies* 5 (2):201–2.
Agamben, Giorgio. 2009. *What is an Apparatus?* Stanford University Press.
Agamben, Giorgio. 2013, 16 November. 'For a theory of destituent power'. Nicolas Poulantzas Institute & SYRIZA Youth, www.chronosmag.eu/index.php/g-agamben-for-a-theory-of-destituent-power.html.
Agencies Mississippi. 2017. 'Mockingbird removed from reading list'. *Guardian*, 16 October, p.13.
Agier, Michel. 2011. *Managing the Undesirables: Refugee Camps and Humanitarian Governance*. Cambridge & Malden, MA: Polity.
AI. 2007. 'Eyes on Darfur'. Amnesty International, www.eyesondarfur.org/index.html.
Alcock, Rupert. 2016. 'Politics and the new unconscious: thinking beyond biopolitics'. Ph.D. thesis, School of Sociology, Politics and International Studies University of Bristol.

BIBLIOGRAPHY

ALNAP. November 2009. '25th ALNAP Annual Meeting: Innovations Fair'. Active Learning Network for Accountability and Performance, www.alnap.org/ourwork/innovations/fair.
Althusser, Louis, and Etienne Balibar. 1970 [1968]. *Reading Capital*. London: New Left Books.
Amoore, Louise. 2011. 'Data derivatives: on the emergence of a security risk calculus for our times'. *Theory, Culture & Society* 28 (6):24–43.
Amsden, A. H. 1990. 'Third World industrialization: "global Fordism" or a new model?' *New Left Review* (182):5–31.
Anderson, Chris. 2007. 'The end of theory: the data deluge makes the scientific method obsolete'. *Wired Magazine*, http://archive.wired.com/science/discoveries/magazine/16-07/pb_theory.
Andreotti, Libero. 2014. 'Unfaithful reflections: re-actualising Benjamin's aestheticisation thesis'. In *Architecture Against the Post-Political: Essays in Reclaiming the Critical Project*, ed. Nadir Lahiji, 41–65. London: Routledge.
Arendt, Hannah. 1994 [1951]. *The Origins of Totalitarianism*. New York: Harcourt, Inc.
Arendt, Hannah. 1998 [1958]. *The Human Condition*. University of Chicago Press.
Asad, Talal. 1973. 'Two European images of non-European rule'. In *Anthropology and the Colonial Encounter*, ed. Talal Asad, 103–18. London: Ithaca Press.
Associated Press. 2017, 16 November. 'Facebook drone that could bring global internet access completes test flight'. *Guardian*, https://www.theguardian.com/technology/2017/jul/02/facebook-drone-aquila-internet-test-flight-arizona.
Balakrishnan, Gopal. 2009. 'Speculations on the stationary state'. *New Left Review* 59:5–26.
Balibar, Étienne 2016. 'Critique in the 21st century: political economy still, and religion again'. *Radical Philosophy* (200):11–21.
Bally, Philippe, Jerome Bequignon, Olivier Arino and Stephen Briggs. 2005. 'Remote sensing and humanitarian aid: a life-saving combination'. *esa bulletin* 122:36–41.
Banaji, Jairus. 1970. 'The crisis of British anthropology'. *New Left Review* (64):71–85.
Bannaga, Sharaf Eldin Ibrahim. 2002. *Peace and the Displaced in the Sudan: The Khartoum Experience*. Zurich: Habitat Group, Swiss Federal Institute of Technology, School of Architecture.
Barbrook, Richard. 2013. *Imaginary Futures: From Thinking Machines to the Global Village*. London: Pluto Press.
Barbrook, Richard, and Andy Cameron. 1995. 'The Californian ideology'. The Hypermedia Research Centre, www.hrc.wmin.ac.uk/theory-californianideology-main.html.
Barnett, Anthony. 1968. 'A revolutionary student movement'. *New Left Review* 1 (53):43–53.
Barnett, Michael. 2011. *Empire of Humanity: A History of Humanitarianism*. Ithaca & London: Cornell University Press.
Bartlett, Jamie. 6 August 2017. *Secrets of Silicon Valley* – series 1:1 'The disruptors' [TV broadcast, 60 minutes]. BBC2, www.bbc.co.uk/programmes/b0916ghq.
Batty, David. 8 April 2008. 'UN uses Google to pinpoint refugee crises'. *Guardian*, www.theguardian.com/world/2008/apr/08/unitednations.sudan.

Beck, Ulrich. 1992 [1986]. *Risk Society: Towards a New Modernity*. London: Sage Publications.
Becker, Kristina Flodman. 2004. *The Informal Economy: Fact Finding Study*. Stockholm: Sida.
Beijing Review. 1992. 'New "open belt" forms along border areas'. *Beijing Review* 35 (5–6):5–6.
Bell, Daniel. 2000 [1960]. *The End of Ideology: On the Exhaustion of Political Ideas in the Fifties*. Cambridge, MA: Harvard University Press.
Betts, Alexander, and Louise Bloom. 2014. *Humanitarian Innovation: The State of the Art*, OCHA Policy and Studies Series. New York: United Nations Office for the Coordination of Humanitarian Affairs.
Bjorgo, Einar. 2002. 'Space aid: current and potential uses of satellite imagery in UN humanitarian organisations'. United States Institute of Peace, http://webharvest.gov/peth04/20041017064049/http://www.usip.org/virtualdiplomacy/publications/reports/12.html.
Black, Maggie. 1992. *A Cause for Our Time: Oxfam – The First 50 Years*. Oxford: Oxfam and Oxford University Press.
Black, Robert E., Lindsay H. Allen, Zuliqar A. Bhutta, et al. 2008. 'Maternal and child undernutrition: global and regional exposures and health consequences'. *Lancet* (371):243–60.
Blackburn, Robin, ed. 1973. *Ideology in Social Science: Reading in Critical Social Theory*. New York: Vintage Books.
Blanchetiere, Pascale. 2006. 'Resilience of humanitarian workers'. People in Aid, www.peopleinaid.org/pool/files/pubs/resilience-of-aid-workers-article.pdf.
Boltanski, Luc, and Eve Chiapello. 2005. *The New Spirit of Capitalism*, trans. Gregory Elliott. London & New York: Verso. Original French edition, 1999.
Boltanski, Luc, and Arnaud Esquerre. 2016. 'The economic life of things: commodities, collectibles, assets'. *New Left Review* (98):31–54.
Bonneuil, Christophe, and Jean-Baptiste Fressoz. 2016. *The Shock of the Anthropocene: The Earth, History and Us*. London & New York: Verso.
Booth, David. 1985. 'Marxism and development sociology: interpreting the impasse'. *World Development* 13 (7):761–87.
Bostdorff, Denise M. 2008. *Proclaiming the Truman Doctrine: The Cold War Call to Arms*. Station: Texas A&M University Press.
Bouchardy, Jean-Yves. 1995. *Development of a GIS System in UNHCR for Environmental, Emergency, Logistic and Planning Purposes*. Geneva: Office of the Senior Coordinator on Environmental Affairs, UNHCR.
Boutang, Yann Moulier. 2011 [2008]. *Cognitive Capitalism*, trans. Ed Emery. Cambridge: Polity.
Boutros-Ghali, Boutros. 1995. 'An agenda for peace: preventive diplomacy, peacemaking and peace-keeping' [17 June 1992], 39–72. In *An Agenda for Peace: 1995*. New York: United Nations.
Bradbury, Mark, Nicholas Leader, and Kate Mackintosh. 2000. *The 'Agreement on Ground Rules' in South Sudan*. London: Humanitarian Policy Group.
Braidotti, Rosi. 2013. *The Posthuman*. Cambridge: Polity.
Briend, André, and Collins Steve. 2010. 'Therapeutic nutrition for children with severe acute malnutrition: summary of African experience'. *Indian Paediatrics* 47 (8):655–9.

Brown, Wendy. 2010. *Walled States, Waning Sovereignty*. New York: Zone Books.
Bruderlein, Claude, and Pierre Gassmann. 2006. 'Managing security risks in hazardous missions: the challenges of securing United Nations access to vulnerable groups'. *Harvard Human Rights Journal* 19:63–93.
Bryant, Raymond L. 2002. 'Non-governmental organizations and governmentality: "consuming" biodiversity and indigenous people in the Philippines'. *Political Studies* 50:268–92.
Brynjolfsson, Erik, and Andrew McAfee. 2011. *Race Against the Machine: How the Digital Revolution is Accelerating Innovation, Driving Productivity, and Irreversibly Transforming Employment and the Economy*. Lexington, MA: Digital Frontier Press.
Buchanan, Keith. 1963. 'The Third World – its emergence and contours'. *New Left Review* 1 (18):5–23.
Buchanan, Keith. 1973. 'The role of the intellectual in aiding the liberation struggle'. *Journal of Contemporary Asia* 3 (1):34–8.
Buck-Morss, Susan. 1977. *The Origin of Negative Dialectics: Theodor W. Adorno, Walter Benjamin, and the Frankfurt Institute*. New York: Free Press.
Burns, Ryan. 2014, 9 October. 'Rethinking big data in digital humanitarianism: practices, epistemologies, and social relations'. *Geojournal*, https://pdfs.semanticscholar.org/4e82/67f9df52d83521c42012735e42a496b337a3.pdf.
Bush, Ray. 2007. *Neoliberalism and Poverty*. London & Ann Arbor, MI: Pluto Press.
Butler, Genevieve. 2003, 17 September. 'PIA code prioritises aid worker safety, but raises costs'. Relief Web, https://reliefweb.int/report/afghanistan/pia-code-prioritises-aid-worker-safety-raises-costs.
CableFree. 2017. 'FSO (Free Space Optics, laser, optical wireless) guide'. CableFree, www.cablefree.net/wirelesstechnology/free-space-optics/fso-guide.
Cadwalladr, Carole. 2017, 26 February. 'Robert Mercer: the big data billionaire waging war on mainstream media'. *Observer*, https://www.theguardian.com/politics/2017/feb/26/robert-mercer-breitbart-war-on-media-steve-bannon-donald-trump-nigel-farage.
Carr, Nicholas. 2015. *The Glass Cage: Where Automation is Taking Us*. London: The Bodley Head.
Castells, Manuel. 1996. *The Rise of the Network Society*. Oxford: Blackwell Publishers Ltd.
Cater, Nick. 1986. *Sudan the Roots of Famine: A Report for Oxfam*. Oxford: Oxfam.
Cederstrom, Carl, and Peter Fleming. 2012. *Dead Man Working*. Winchester & Washington: Zone Books.
Chamayou, Gregoire. 2015 [2013]. *Drone Theory*, trans. Janet Lloyd. London: Penguin.
Chambers, Robert. 1983. *Rural Development: Putting the Last First*. London: Longman.
Chandler, David. 2013. '"Human-centred" development? Rethinking "freedom" and "agency" in discourses of international development'. *Millennium: Journal of International Studies* 42 (1):3–23.

Chandler, David. 2015. 'A world without causation: big data and the coming of the age of posthumanism'. *Millennium: Journal of International Studies* 43 (3):833–51.

Chandler, David. 2016a. 'How the world learned to stop worrying and love failure: big data, resilience and emergent causality'. *Millennium: Journal of International Studies* 44 (3):391–410.

Chandler, David. 2016b. 'New narratives of international security governance: the shift from the global interventionism to global self-policing'. *Global Crime* 17 (3–4):262–80.

Chandler, David. 2018. *Ontopolitics in the Anthropocene: An Introduction of Mapping, Sensing and Hacking*. London: Routledge

Chouliaraki, Lilie. 2013. *The Ironic Spectator: Solidarity in the Age of Post-Humanitarianism*. Cambridge & Malden, MA: Polity.

Clark, Tom, and Anthony Heath. 2015. *Hard Times: Inequality, Recession, Aftermath*. New Haven, CT, & London: Yale University Press.

Collier, Stephen J., and Andrew Lakoff. 2007. 'Vital systems security'. ARC Working Paper No. 2, www.anthropos-lab.net/workingpapers/no2.pdf.

Collins, Philip. 2016, 23 December. 'Never forget that we live in the best of times'. *The Times*, https://www.thetimes.co.uk/article/never-forget-that-we-live-in-the-best-of-times-5n97c9xs0.

Collinson, Sarah, and Mark Duffield, with Carol Berger, Diana Felix da Costa, and Karl Sandrom. 2013. *Paradoxes of Presence: Risk Management and Aid Culture in Challenging Environments*. London: Humanitarian Policy Group (HPG), Overseas Development Institute (ODI).

Collinson, Sarah, and Samir Elhawary. 2012. *Humanitarian Space: A Review of Trends and Issues*, HPG Report No. 32. London: Humanitarian Policy Group, Overseas Development Institute.

Comoretto, Amanda, Nicola Crichton, and Ian Albery. 2011. *Resilience in Humanitarian Aid Workers: Understanding Processes of Development*. London: Lambert Academic Publishing.

Conneally, Paul. November 2011. 'Digital humanitarianism' [video (10 minutes 57 seconds)], www.ted.com/talks/paul_conneally_digital_humanitarianism.

Conrad, Joseph. 2002. *The Heart of Darkness and Other Tales*. Oxford University Press.

Coole, Diana. 2013. 'Agentic capacities and capacious historical materialism: thinking with new materialisms in the political sciences'. *Millennium: Journal of International Studies* 41 (3):451–69.

Cooper, Melinda. 2008. *Life as Surplus: Biotechnology and Capitalism in the Neoliberal Era*. Seattle & London: University of Washington Press.

Cooper, Melinda. 2011. 'Complexity theory after the financial crisis: the death of neoliberalism or the triumph of Hayek?' *Journal of Cultural Economy* 4 (4):371–85.

Corlett, Adam. 2017. *As Time Goes By: Shifting Incomes and Inequality between and within Generations*. London: Intergenerational Commission.

Cornia, Giovanni Andrea, ed. 1987. 'Economic decline and human welfare in the first half of the 1980s'. In *Adjustment with a Human Face*, volume I, ed. G. A. Cornia, R. Jolly and F. Stewart, 11–47. Oxford: Clarendon Press.

Cortada, James W. 2012. *The Digital Flood: The Diffusion of Information*

Technology Across the US, Europe, and Asia. Oxford & New York: Oxford University Press.
Coward, Martin. 2007. '"Urbicide" reconsidered'. *Theory and Event* 10 (2):paras. 1–56.
Cowen, Deborah. 2014. *The Deadly Life of Logistics: Mapping Violence in Global Trade*. Minneapolis & London: University of Minnesota Press.
Cowen, Harry. 1999. *Community Care, Ideology and Social Policy*. London: Prentice Hall Europe.
Crary, Jonathan. 2014. *24/7: Late Capitalism and the End of Sleep*. London & New York: Verso.
Cross, Jamie. 2013. 'The 100th object: solar lighting technology and humanitarian goods'. *Journal of Material Culture* 18 (4):367–87.
Cross, Jamie, and Alice Street. 2009. 'Anthropology at the bottom of the pyramid'. *Anthropology Today* 25 (4):4–9.
Crowe, Anna. 2013, 11 December. 'United Nations' drones: a sign of what's to come?' Privacy International, https://www.privacyinternational.org/blog/united-nations-drones-a-sign-of-whats-to-come.
Crowley, John, and Jennifer Chan. 2011. *Disaster Relief 2.0: The Future of Information Sharing in Humanitarian Emergencies*. Washington DC and Berkshire, UK: Harvard Humanitarian Initiative for the UN Foundation, UNOCHA & the Vodafone Foundation.
Crutcher, Michael, and Matthew Zook. 2009. 'Placemarks and waterlines: racialized cyberspaces in post-Katrina Google Earth'. *Geoforum* 40:523–34.
Cuppens, Yvonne. 1998. *Research into Food Security in Maganja da Costa, Zambezia, Mozambique*. Maputo: Action Aid Mozambique.
Cutler, Peter. 1984a. 'Famine forecasting: prices and peasant behaviour in Northern Ethiopia'. *Disasters* 8 (1):48–56.
Cutler, Peter. 1984b. 'Food crisis detection: going beyond the balance sheet'. *Food Policy* 9 (3):189–92.
Cutler, Peter. 1987. 'Early warning of famine: a red herring?' *Proceedings of the Nutritional Society* 46:263–6.
Cutts, Mark, and Alan Dingle. 1995. *Safety First: Protecting NGO Employees Who Work in Areas of Conflict*. London: Save the Children Fund.
da Costa, Diana Felix. 2012, December. 'Working in challenging environments: risk management and aid culture in South Sudan'. Global Insecurities Centre, www.bristol.ac.uk/global-insecurities/esrc-dfid/reports/felix.pdf.
Daly, M W., and Jane R. Hogan. 2005. *Images of Empire: Photographic Sources for the British in the Sudan*. Leiden & Boston: Brill.
Davis, Mike. 2006. *Planet of Slums*. London & New York: Verso.
Davis, Mike, and Daniel Bertrand Monk, eds. 2007. *Evil Paradises: Dreamworlds of Neoliberalism*. New York & London: The New Press.
de Bruijin, Mirjam, Francis Nyamnjoh, and Inge Brinkman, eds. 2009. *Mobile Phones: The New Talking Drums of Everyday Africa*. Bamenda, Cameroon & Leiden: Langaa Research Group & Africa Studies Centre.
de Waal, Alex. 1989. *Famine that Kills: Darfur, Sudan, 1984–85*. Oxford: Clarendon Press.
Dean, Mitchell. 1999. *Governmentality: Power and Rule in Modern Society*. London: Sage Publications Ltd.

Deleuze, Gilles. 1992. 'Postscript on the societies of control'. *October* 59:3–7.
Development Initiatives. 2017. 'Global humanitarian assistance report 2017'. Bristol: Development Initiatives
DFID. 2011. *Defining Disaster Resilience: A DFID Approach Paper*. London: Department for International Development.
DFID. 2012. *Promoting Innovation and Evidence-Based Approaches to Building Resilience and Responding to Humanitarian Crises: A DFID Strategy Paper*. London: Department for International Development.
Dillon, Michael. 2007. 'Governing through contingency: the security of biopolitical governance'. *Political Geography* 26 (1):41–7.
Dillon, Michael, and Julian Reid. 2000. 'Global governance, liberal peace and complex emergency'. *Alternatives* 25 (1):117–43.
Dillon, Michael, and Julian Reid. 2009. *The Liberal Way of War: Killing to Make Life Live*. Abingdon: Routledge.
Dobbs, Leo. 8 April 2008. 'UNHCR and Google Earth unveil programme for humanitarian operations'. UNHCR, www.unhcr.org/news/NEWS/47fb8b5b2.html.
Donini, Antonio. 2009. *Afghanistan: Humanitarianism under Threat*. Boston: Feinstein International Center, Tufts University.
Donnell, Hayden. 2017, 29 January. 'Silicon Valley super-rich head south to escape from a global apocalypse'. *Observer*, https://www.theguardian.com/technology/2017/jan/29/silicon-valley-new-zealand-apocalypse-escape.
Donnelly, Rich. 2012. 'Crisis mapping: a multidisciplinary effort'. SPIE Professional July 2012, http://spie.org/x87811.xml.
Donovan, Kevin P. 2013. 'Infrastructuring aid: materializing social protection in Northern Kenya'. CSSR Working Paper No 333. University of Cape Town: Centre for Social Science Research.
Duffield, Mark. 1981. *Maiurno: Capitalism and Rural Life in Sudan*. London: Ithaca Press.
Duffield, Mark. 1983. 'Change among West African settlers in Northern Sudan'. *Review of African Political Economy* 10 (26):45–59.
Duffield, Mark. 1984. 'New racism ... new realism: two sides of the same coin'. *Radical Philosophy* (37):29–34.
Duffield, Mark. 1988. *Black Radicalism and the Politics of De-industrialisation: The Hidden History of Indian Foundry Workers*. Aldershot: Avebury.
Duffield, Mark. 1994. 'Complex emergencies and the crisis of developmentalism'. *Institute of Development Studies Bulletin: Linking Relief and Development* 25:37–45.
Duffield, Mark. 1996. 'Social reconstruction in Croatia and Bosnia: an exploratory report for SIDA'. Stockholm: Swedish International Development and Co-operation Agency (SIDA).
Duffield, Mark. 2001. *Global Governance and the New Wars: The Merger of Development and Security*. London: Zed Books.
Duffield, Mark. 2007. *Development, Security and Unending War: Governing the World of Peoples*. Cambridge: Polity.
Duffield, Mark. 2010. 'The fortified aid compound: everyday life in post-interventionary society'. *Journal of Intervention and Statebuilding* 4 (4):453–74.

Duffield, Mark. 2011. 'Juba Report'. Global Insecurities Centre, www.bristol. ac.uk/media-library/sites/global-insecurities/migrated/documents/juba.pdf.
Duffield, Mark. 2013. 'Disaster-resilience in the network age: access-denial and the rise of cyber-humanitarianism'. DIIS Working Paper 2013:23. Copenhagen: Danish Institute of International Studies (DIIS).
Duffield, Mark. 2014. 'From immersion to simulation: remote methodologies and the decline of area studies'. *Review of African Political Economy* 41 (supp. 1):S75–S94.
Duffield, Mark, Khassim Diagne, and Vicky Tennant. 2008. *Evaluation of UNHCR's Returnee Reintegration Programme in Southern Sudan*. Geneva: Policy Development and Evaluation Service (PDES), United Nations High Commissioner for Refugees.
Duffield, Mark, and Vernon Hewitt. 2009. 'Liberal interventionism and fragile states: linked by design?' In *Empire, Development and Colonialism: The Past in the Present*, ed. Mark Duffield and Vernon Hewitt, 116–29. Oxford & Rochester: James Currey & Boydell and Brewer Inc.
Duffield, Mark, and John Prendergast. 1994. *Without Troops and Tanks: Humanitarian Intervention in Eritrea and Ethiopia*. Trenton, NJ: Red Sea Press / Africa World Press Inc.
Duffield, Mark, Helen Young, John Ryle and Ian Henderson. 1995. *Sudan Emergency Operations Consortium (SEOC): A Review*. Birmingham: School of Public Policy.
Dumont, René, and Nicholas Cohen. 1980. *The Growth of Hunger: A New Politics of Hunger*. London: Marion Boyers Publishing Ltd.
Dupuy, Jean-Pierre. 2000. *The Mechanization of the Mind: On the Origins of Cognitive Science*. Princeton University Press.
Eagle, Nathan, and Alex Pentland. 2006. 'Reality mining: sensing complex social systems'. *Personal Ubiquitous Computing* (10):255–68.
Easterling, Keller. 2005. *Enduring Innocence*. Cambridge, MA: The MIT Press.
Easterling, Keller. 2014. *Extrastatecraft: The Power of Infrastructure Space*. London & New York: Verso.
EC. 1996. *Linking Relief, Rehabilitation and Development (LRRD)*. Brussels: Commission of the European Communities.
ECHO. 2004. *Generic Security Guide for Humanitarian Organisations*. Brussels: European Commission Humanitarian Office.
Edkins, Jenny. 2000. *Whose Hunger? Concepts of Famine, Practices of Aid*. Minneapolis: University of Minnesota Press.
Edwards, Michael. 1989. 'The irrelevance of development studies'. *Third World Quarterly* 11 (1):116–35.
Edwards, Paul N. 2010. *A Vast Machine: Computer Models, Climate Data, and the Politics of Global Warming*. Cambridge, MA: Massachusetts Institute of Technology (MIT).
Eide, Espen Barth, Anja Therese Kaspersen, Randolph Kent and Karen von Hippel. 2005. 'Report on integrated missions: practical perspectives and recommendations'. UN ETCHA Core Group.
Elden, Stuart. 2013. 'Secure the volume: vertical geopolitics and the depth of power'. *Political Geography* 34:35–51.

Eldredge, Elizabeth, and Denis Rydjeski. 1988. 'Food crises, crisis response and emergency preparedness'. *Disasters* 12 (1):1–5.

Eldredge, Elizabeth, Cordella Salter and Denis Rydjeski. 1986. 'Towards an early warning system in Sudan'. *Disasters* 10 (3):189–96.

EnviRef. 2001. 'Environmental monitoring of refugee camps using high-resolution satellite images: final report'. Brussels European Commission DGXII Environment and Climate Program in partnership with NERSC, Norway; Satellus, Sweden; Infocato, Spain; and UNHCR, Geneva.

Errington, Fredrick, Tatsuro Fujikura and Deborah Gewertz. 2012. 'Instant noodles as an antifriction device: making the BOP with PPP in PNG'. *American Anthropologist* 114 (1):19–31.

Evans, Brad, and Julian Reid. 2014. *Resilient Life: The Art of Living Dangerously*. Cambridge: Polity.

Faris, James. 1973. 'Pax Britannica and the Sudan: S. F. Nadel'. In *Anthropology and the Colonial Encounter*, ed. Talal Asad, 153–72. London: Ithaca Press.

FindBiometrics. 2016, 1 March. 'Crossmatch biometrics tech to register migrants in Greece'. FindBiometrics, https://findbiometrics.com/crossmatch-biometrics-tech-to-id-and-register-migrants-in-greece-303011.

Fleming, Sue, and Colin Barnes. 1992. *Poverty Options in Mozambique: Strategy Options for Future Aid*. University of Manchester: Department of Social Anthropology.

Foer, Franklin. 2017. *World Without Mind: The Existential Threat of Big Tech*. London: Jonathan Cape.

Folke, Carl. 2006. 'Resilience: the emergence of a perspective for social-ecological systems analysis'. *Global Environmental Change* 16 (3):253–67.

Foucault, Michel. 1998 [1976]. *The Will to Knowledge: The History of Sexuality*, volume I. London: Penguin Books.

Foucault, Michel. 2003. *Society Must be Defended: Lectures at the Collège de France, 1975–76*. London: Allen Lane, The Penguin Press.

Foucault, Michel. 2007. *Security, Territory, Population: Lectures at the Collège de France, 1977–1978*. Basingstoke: Palgrave Macmillan.

Foucault, Michael. 2013. *Freedom and Knowledge: A Hitherto Unpublished Interview*, ed. Fons Elders. Amsterdam: Elders Special Productions BV.

Fox, Fionna. 1999. 'The politicisation of humanitarian aid: a discussion paper for Caritas Europa – November 1999'. London: CAFOD.

Franklin, Victor. 2016, 31 December. 'Bathrooms are coming: an internal history of corporate comms' [Archive on 4 podcast (58 minutes)]. BBC Radio 4, www.bbc.co.uk/programmes/b086knhm.

Fraser, Nancy. 2012, 23 August. 'Can society be commodities all the way down? Polanyian reflections on capitalist crisis'. FMSH-WP-2012-18 <halshs-00725960>, https://halshs.archives-ouvertes.fr/halshs-00725060/document.

Fraser, Nancy. 2016. 'Contradictions of capital and care'. *New Left Review* (100):99–117.

Fraser, Nancy. 2017, 2 January. 'The end of progressive neoliberalism'. *Dissent*, https://www.dissentmagazine.org/online_articles/progressive-neoliberalism-reactionary-populism-nancy-fraser.

Fredriksen, Aurora. 2014. 'Emergency shelter topologies: locating humanitarian space in mobile and material practice'. *Environment and Planning D: Society and Space* 31:147–62.

Friedman, Eli D., and Ching Kwan Lee. 2010. 'Remaking the world of Chinese labour: a 30-year retrospective'. DigitalCommons@ILR. http://digitalcommons.ilr.cornell.edu/cgi/viewcontent.cgi?article=1845&context=articles.

Fukuyama, Francis. 2002 [1989]. 'The end of history?' In *Globalization and the Challenges of a New Century*, ed. Patrick O'Meara, Howard D. Mehlinger and Matthew Krain, 161–80. Bloomington and Indianapolis: Indiana University Press.

Furedi, Frank. 1994. 'The new Crusades'. *Living Marxism* 65:18–21.

Gabrys, Jennifer. 2016. *Program Earth: Environmental Sensing Technology and the Making of a Computational Planet*. Minneapolis: University of Minesota Press.

Galloway, Alexander R. 2013. 'The poverty of philosophy: realism and post-Fordism'. *Critical Enquiry* 39 (2):347–66.

Geldof, Bob. 1985. 'Sayings of the week'. *Observer*, 27 October.

George, Susan. 1976. *How the Other Half Dies: The Real Reasons for World Hunger*. Harmondsworth: Penguin Books.

Geotz, Thomas. 2011, 19 June. 'Harnessing the power of feedback loops'. *Wired*, https://www.wired.com/2011/06/ff_feedbackloop.

Gibson, William. 1984. *Neuromancer*. New York: Ace Books.

Gillula, Jeremy, and Jeremy Malcolm. 2015, 18 May. 'Internet.org is not neutral, not secure, and not the Internet'. Electronic Frontier Foundation, https://www.eff.org/deeplinks/2015/05/internetorg-not-neutral-not-secure-and-not-internet.

Giroux, Jennifer. 2009. 'We all have a role to play: the role of society in preparing for responding to emergencies'. *FrontLine Security* 4 (2):24–7.

Gonzalez, Roberto J. 2008. '"Human terrain": past, present and future applications'. *Anthropology Today* 24 (1):21–6.

Goodhart, David. 2017. *The Road to Somewhere: The Populist Revolt and the Future of Politics*. London: Hurst.

Google. 2015. 'Project Loon: balloon-powered Internet for everyone'. Google, www.google.com/loon.

Gorz, Andre. 1982. *Farewell to the Working Class*. London: Pluto.

Graham, Stephen. 2013, 14 February. 'Foucault's boomerang: the new military urbanism'. OpenSecurity: Conflict and Peacebuilding, https://www.opendemocracy.net/opensecurity/stephen-graham/foucault%e2%80%99s-boomerang-new-military-urbanism.

Graham, Stephen, and Simon Marvin. 2001. *Splintering Urbanism*. London: Routledge.

Green, Duncan. 2014. 'The new World Development Report (on mind, society and behaviour): lots to like, but a big fail on power, politics and religion'. Oxfam, Last Modified 16 December 2017, http://oxfamblogs.org/fp2p/lots-to-like-in-the-new-world-development-report-on-mind-society-and-behavior-but-a-big-fail-on-power-politics-and-religion.

Gregory, Derek. 2008. 'The biopolitics of Baghdad: counterinsurgency and the counter-city'. Human Geography, http://web.mac.com/derekgregory/iWeb/Site/The%20biopolitics%20of%20Baghdad.htm.

Grinstead, Nick. 2016, 22 February. 'The Khartoum process: shifting the burden'.

Clingendael (Netherlands Institute of International Relations), https://www.clingendael.nl/publication/khartoum-process-shifting-burden.
Haidt, Jonathan, and Nick Haslam. 2016, 10 April. 'Campuses are places for open minds – not where debate is closed down'. *Guardian*, www.theguardian.com/commentisfree/2016/apr/10/students-censorship-safe-places-platforming-free-speech.
Halpern, Orit. 2014a. *Beautiful Data: A History of Vision and Reason since 1945*. Durham & London: Duke University Press.
Halpern, Orit. 2014b. 'Cybernetic rationality.' *Distinktion: Scandinavian Journal of Social Theory* 15 (2):223–38.
Halpern, Orit. 2014c. 'Inhuman vision'. *Media-N: Journal of the New Media Caucus* – Special Issue on Infrastructure and Art, http://median.newmediacaucus.org/art-infrastructures-information/inhuman-vision.
Halpern, Orit. 2017, April. 'Hopeful resilience'. e-flux architecture, www.e-flux.com/architecture/accumulation/96421/hopeful-resilience.
Hammerstad, Anne. 2014. *The Rise and Decline of a Global Security Actor: UNHCR, Refugee Protection, & Security*. Oxford University Press
Hanchard, Doug. 2012. *Constructive Convergence: Imagery and Humanitarian Assistance*. Washington DC: National Defence University, Institute for National Strategic Studies, Centre for Technology and National Security Policy.
Harman, Graham. 2010. *Towards Speculative Realism*. Winchester, UK, & Washington, DC: Zero Books.
Harrell-Bond, Barbara. 1998. 'Camps: literature review'. *Forced Migration Review* 2:22–3.
Harris, Chad. 2006. 'The omniscient eye: satellite imagery, "battlespace awareness", and the structures of imperial gaze'. *Surveillance and Society* 4 (1–2):101–22.
Hayek, F. A. 1945. 'The use of knowledge in society'. *The American Economic Review* 35 (4):519–32.
Hayes, Tim. 2012. 'DSS 2012: bringing new IR sensors to market – changes in military needs mean both challenges and opportunities for new IR sensing technologies'. optics.org, http://optics.org/news/3/4/40.
Hayles, Katherine. 1999. *How We Became Posthuman: Virtual Bodies in Cybernetics, Literature, and Informatics*. Chicago & London: University of Chicago Press.
Headrick, Daniel R. 2012. *The Invisible Weapon: Telecommunications and International Politics, 1851–1945*, reprint edn. Oxford & New York: Oxford University Press.
Healy, Sean, and Sandrine Tiller. 2014. *Where is Everyone? Responding to Emergencies in the Most Difficult Places*. London: Medecins Sans Frontières (MSF) UK.
HERR. 2011. *Humanitarian Emergency Response Review*. London: Commissioned by Department for International Development (DFID).
Hewitt, Kenneth. 1983. 'The idea of calamity in a technocratic age'. In *Interpretations of Calamity from the Viewpoint of Human Ecology*, ed. Kenneth Hewitt, 1–32. Boston, London & Sydney: Allen & Unwin Inc.
Hirst, Paul, and Grahame Thompson. 1996. *Globalisation in Question*. Cambridge: Polity.

Hobsbawm, Eric. 1994. *The Age of Extremes: The Short Twentieth Century*. London: Michael Joseph.
Holling, Crawford S. 1973. 'Resilience and stability of ecological systems'. *Annual Review of Ecology and Systematics* 4:1–23.
Homer-Dixon, Thomas. 2007. *The Upside of Down: Catastrophe, Creativity, and the Renewal of Civilisation*. London: Souvenir Press Ltd.
Honan, Mat. 2014, 24 February. 'Facebook's plan to conquer the world – with crappy phones and bad networks'. *Wired*, www.wired.com/gadgetlab/2014/02/facebook-plans-conquer-world-slew-low-end-handsets.
Horkheimer, Max, and Theodor W. Adorno. 1979. *Dialectic of Enlightenment*. London: Verso.
Hosein, Gus, and Carly Nyst. 2013. *Aiding Surveillance: An Exploration of How Development and Humanitarian Aid Initiatives Are Enabling Surveillance in Developing Countries*. London: Privacy International.
Howell, Alison. 2012. 'The imminent demise of PTSD: from governing trauma to governance through resilience'. *Alternatives: Global, Local, Political* 36 (2):36–49.
HRE. 2009. 'Darfur is dying' [online computer game]. Human Rights Education, UN Regional Information Centre for Western Europe (UNRIC), developed by Reebok Human Rights Foundation, International Crisis Croup, mtvU and students at University of Southern California, www.humanrightseducation.info/game-darfur-is-dying.html.
Huws, Ursula. 2015, 10 May. 'Labour in the global digital economy: cybertariat comes of age' [video]. Academy of Fine Arts, https://www.youtube.com/watch?v=dG-xj42eZW8.
Iazzolino, Gianluca. 2015. 'Following mobile money in Somaliland'. Rift Valley Institute Research Paper No. 4. London: Rift Valley Institute (RVI).
IFRC. 2013. *World Disasters Report: Focus on Technology and the Future of Humanitarian Intervention*. Geneva: International Federation of Red Cross and Red Crescent Societies.
IRIN. 2010. 'Health: aiding aid workers'. *Humanitarian News and Analysis* (UN Office for the Coordination of Humanitarian Affairs).
Jackson, Terence. 2016, 21 January. 'Why the voice of Africa's informal economy should be heard'. The Conversation, https://theconversation.com/why-the-voice-of-africas-informal-economy-should-be-heard-52766.
Jacobsen, Katja Lindskov. 2015. *The Politics of Humanitarian Technology: Good Intentions, Unintended Consequences and Insecurity*. London & New York: Routledge.
James, C. L. R. 2001 [1938]. *The Black Jacobins: Toussaint L'Ouverture and the San Domingo Revolution*. London: Penguin Books.
Jarmolowski, Maggie. 21 January 2012. 'Psychological support during SBTF crisis mapping deployments'. Standby Task Force, http://blog.standbytaskforce.com/2012/01/21/psychological-support-during-sbtf-crisis-mapping-deployments.
Jaspars, Susanne. 2015. 'Food aid, power and profit: an historical analysis of the relation between food aid and governance in Sudan'. Ph.D. thesis, School of Politics and International Studies (SPAIS), University of Bristol.
Jaspars, Susanne, and Jeremy Shoham. 1999. 'Targeting the vulnerable: A review

of the necessity and feasibility of targeting vulnerable households'. *Disasters* 23 (4):359–72.
Johnson, Cedric G. 2011. 'The urban precariat, neoliberalization, and the soft power of humanitarian design'. *Journal of Developing Societies* 27 (3&4):445–75.
Johnson, Douglas H. 1982. 'Evans-Pritchard, the Nuer and the Sudan Political Service'. *African Affairs* 81 (323):231–46.
Jones, Owen. 2011. *Chavs: The Demonization of the Working Class*. London & New York: Verso.
Joseph, Jonathan. 2016. 'Governing through failure and denial: the new resilience agenda'. *Millennium: Journal of International Studies* 44 (3):370–90.
Joshi, Dhaval. 2017. 'Why robots will kill middle incomes'. European Investment Strategy: Special Report August 10, 2017. Montreal & London: BCA Research
Jutting, Johannes P., and Juan R. de Laiglesia, eds. 2009. *Is Informal Normal? Towards More and Better Jobs in Developing Countries*. Paris: Development Centre of the Organisation for Economic Cooperation and Development (OECD).
Kahn, Clea. 2008. *Conflict, Arms, and Militarization: The Dynamics of Darfur's IDP Camps*. Geneva: Small Arms Survey, Graduate Institute of International and Development Studies.
Kaldor, Mary. 1999. *New and Old Wars: Organised Violence in a Global Era*. Cambridge: Polity.
Kaplan, Robert D. 1994. 'The coming anarchy: how scarcity, crime, overpopulation, and disease are rapidly destroying the social fabric of our planet'. *Atlantic Monthly*: 44–76.
Karadawi, Ahmed. 1999. *Refugee Policy in Sudan: 1967–1984*. New York: Berghahn Books.
Karim, Ataul, Mark Duffield, Susanne Jaspars, et al. 1996. *Operation Lifeline Sudan (OLS): A Review*. Geneva: Department of Humanitarian Affairs.
Keen, David. 1994. *The Benefits of Famine: A Political Economy of Famine and Relief in Southwestern Sudan, 1983–1989*. Princeton University Press.
Kemper, Thomas, Malgorzata Jenerowicz, Martino Pesaresi and Pierre Soille. 2011. 'Enumeration of dwellings in Darfur camps from GeoEye-1 satellite images using mathematical morphology'. *IEEE Journal of Selected Topics in Applied Earth Observation and Remote Sensing* 4 (1):8–15.
Kibera. 2017. 'Kibera facts & information'. Kibera UK, www.kibera.org.uk/facts-info.
Klein, Naomi. 2007. *The Shock Doctrine: The Rise of Disaster Capitalism*. London: Penguin Books Ltd.
Kranz, Olaf, Veronica Gstaiger, Stefan Lang, et al. 2010. 'Different approaches for IDP camp analyses in West Darfur (Sudan) – a status report'. Politecnico di Torino, http://publications.jrc.ec.europa.eu/repository/handle/111111111/16251.
Kristof, Nicholas. 21 January 2017. 'Why 2017 may be the best year ever'. *New York Times*, https://www.nytimes.com/2017/01/21/opinion/sunday/why-2017-may-be-the-best-year-ever.html.
Labour Party. 1964. *The New Britain*. London: Labour Party.

Lafontaine, Celine. 2007. 'The cybernetic matrix of "French theory"'. *Theory, Culture & Society* 24 (5):27–46.

Lang, Stefan, Dirk Tiede, Daniel Hölbling, Petra Füreder and Peter Zeil. 2010. 'Earth Observation (EO)-based ex post assessment of internally displaced person (IDP) camp evolution and population dynamics in Zam Zam, Darfur'. *International Journal of Remote Sensing* 31 (21):5709–31.

Large, Dan. 2012. 'Fantastic invasions: interventions and the politics of the international in Sudan'. Ph.D. thesis, School of Oriental and African Studies.

Latour, Bruno. 1987. *Science in Action: How to Follow Scientists and Engineers Through Society*. Milton Keynes: Open University Press.

Latour, Bruno. 1992. *We Have Never Been Modern*. Cambridge, MA: Harvard University Press.

Latour, Bruno. 2004. 'Why has critique run out of steam? From matters of fact to matters of concern'. *Critical Enquiry* 30 (2):225–48.

Latour, Bruno. 2008. 'A cautious Prometheus? A few steps towards a philosophy of design (with special attention to Peter Sloterdijk)'. Networks of Design Falmouth, Cornwall, 3 September 2008.

Latour, Bruno, Pablo Jensen, Tommaso Venturini, Sebastian Grauwn and Dominique Boullier. 2012. '"The whole is always smaller than its parts" – a digital test of Gabriel Tarde's monads'. *British Journal of Sociology* 63 (4):590–615.

Lavers, C., C. Bishop, O. Hawkins, et al. 2009. 'Application of satellite imagery to monitoring human rights abuse of vulnerable communities, with minimal risk to relief staff'. *Journal of Physics: Conference Series – Sensors and their Applications XV* 178 (1):1–6.

Lavinas, Lena. 2013. '21st century welfare'. *New Left Review* (84):5–40.

LeBaron, Genevieve, and Alison Ayers. 2013. 'The rise of a "New Slavery"? Understanding unfree labour through neoliberalism'. *Third World Quarterly* 34 (5):837–92.

Lesczynski, Agnieszka. 2012. 'Situating the geoweb in political economy'. *Progress in Human Geography* 36 (1):72–89.

Levine, Iain. 1997. 'Promoting humanitarian principles: the South Sudan experience'. Relief and Rehabilitation Network (RRN), Network Paper: 21. London: Overseas Development Institute.

Lévi-Strauss, Claude. 1966. *The Savage Mind*. London: Weidenfeld & Nicolson.

Lévi-Strauss, Claude. 1968. *Structural Anthropology*. London: Allen Lane, The Penguin Press.

Lewis, Michael. 2014. *Flash Boys: Cracking the Money Code*. London: Penguin.

Light, Jennifer S. 2003. *From Warfare to Welfare: Defence Intellectuals and Urban Problems in Cold War America*. Baltimore, MD: Johns Hopkins University Press.

Lugard, Lord. 1965. *The Dual Mandate in Tropical Africa*. London: Frank Cass.

Lukasik, Stephen J. 2011. 'Why the Arpanet was built'. *IEEE Annals of the History of Computing* 33 (3):4–21.

Luxemburg, Rosa. 1916. 'The Junius Pamphlet – the crisis of German Social Democracy'. In Luxemburg Internet Archive (marxists.org) 2003, https://www.marxists.org/archive/luxemburg/1915/junius/ch01.htm.

Macmichael, H. A. 1934. *The Anglo-Egyptian Sudan*. London: Faber and Faber.

Macrae, Joanna, and Nicholas Leader. 2000. *Shifting Sands: The Search for 'Coherence' Between Political and Humanitarian Responses to Complex Emergencies*. London: Overseas Development Institute.

Mahmoud, Fatima Babiker. 1984. *The Sudanese Bourgeoisie: Vanguard of Development?* London & Khartoum: Zed Books & University of Khartoum.

MapAction. 2008. 'Google Earth and its potential in the humanitarian sector: a briefing paper'.

Marcuse, Herbert. 1967. 'The question of revolution'. *New Left Review* (45):3–7.

Marcuse, Herbert. 1968. *One Dimensional Man: The Ideology of Industrial Society*. London: Sphere Books.

Marcuse, Herbert. 1969. 'Re-examination of the concept of revolution'. *New Left Review* (56):27–34.

Marcuse, Herbert. 1972. *Counterrevolution and Revolt*. London: Allen Lane, The Penguin Press.

McBain, Sophie. 2014, 10 April. 'In Syria, the internet has become just another battleground'. *New Statesman*, https://www.newstatesman.com/politics/2014/04/syria-internet-has-become-just-another-battleground.

McFate, Montgomery. 2004. 'The military utility of understanding adversary culture'. *Joint Forces Quarterly* (38):42–8.

Mckinsey Global Institute. 2016. *Bridging Global Infrastructure Gaps*. Mckinsey&Company.

Meagher, Kate. 1998. *A Back Door to Globalisation? Structural Adjustment, Globalisation and Transborder Trade in West Africa*. University of Oxford, Nuffield College.

Meagher, Kate. 2015. 'Leaving no one behind? Informal economies, economic inclusion and Islamic extremism in Nigeria'. *Journal of International Development* 27:835–55.

Meagher, Kate. 2016. 'The scramble for Africans: demography, globalisation and Africa's informal labour markets'. *Journal of Development Studies* 52 (4):483–97.

Meier, Patrick. 2012, 9 April. 'Does the humanitarian industry have a future in the digital age?' iRevolutions, https://irevolutions.org/tag/organizations.

Meier, Patrick. 2013a, 9 April. 'Humanitarianism in the network age: groundbreaking study'. iRevolutions, http://irevolutions.org/2013/04/09/humanitarianism-network-age/#comment-36134.

Meier, Patrick. 2013b, 11 January. 'How to create resilience through Big Data'. iRevolutions, http://irevolutions.org/2013/01/11/disaster-resilience-2-0.

Meier, Patrick. 2015. *Digital Humanitarians: How BIG DATA is Changing the Face of Humanitarian Response*. Boca Raton, London & New York: CRC Press.

Meyers, Eytan. 2002. 'The causes and convergence in Western immigration control'. *Review of International Studies* 28 (1):123–41.

Mills, C. Wright. 1960. 'Letter to the New Left'. *New Left Review* (5):18–23.

Mills, Mark P. 2013, August. 'The cloud begins with coal: big data, big networks, big infrastructure, and big power – an overview of the electricity used by the global digital ecosystem'. Digital Power Group: Sponsored by National Mining Association & American Coalition for Clean Coal Energy, www.tech-pundit.com/wp-content/uploads/2013/07/Cloud_Begins_With_Coal.pdf?c761ac.

Minton, Anna. 2009. *Ground Control: Fear and Happiness in the Twenty-First-Century City*. Harmondsworth: Penguin Books.
Mishra, Pankaj. 2017a. *Age of Anger: A History of the Present*. London: Allen Lane, Penguin Random House.
Mishra, Pankaj. 2017b. 'What is great about ourselves.' *London Review of Books* 39 (18):3–7.
Mohamed Ahmed Ali, Taisier. 1989. *The Cultivation of Hunger: State and Agriculture in Sudan*. Khartoum University Press.
Montgomery, Charles. 2009. 'The archipelago of fear: are fortifications and foreign aid making Kabul more dangerous?' *The Walrus*, https://thewalrus.ca/the-archipelago-of-fear.
Morozov, Evgeny. 2013. *To Save Everything, Click Here – Technology, Solutionism and the Urge to Fix Problems That Don't Exist*. London: Allen Lane.
Morozov, Evgeny. 2015, 7 June. 'Where Uber and Amazon rule: welcome to the world of the platform'. *Guardian*, www.theguardian.com/technology/2015/jun/07/facebook-uber-amazon-platform-economy.
Morozov, Evgeny. 2017, 22 October. 'Google's plan to revolutionise cities is a takeover in all but name'. *Guardian*, https://www.theguardian.com/technology/2017/oct/21/google-urban-cities-planning-data.
Morrow, Monique. 2016, 22 October. 'Why we need to build a humanized internet'. Linkedin Pulse, https://www.linkedin.com/pulse/why-we-need-build-humanized-internet-monique-morrow.
MSF. 2017. '1971: the creation of Medecins Sans Frontières'. Medecins Sans Frontières, www.msf.fr/histoire-sommaire-book-page/1971-creation-medecins-sans-frontieres.
MSR. 2017. 'Munich Security Report 2017: post-truth, post-West, post-order?' Munich Security Conference.
Muhren, Willem J., and Bartel van de Walle. 2010. 'Sense-making and information management in emergency response'. *Bulletin of the American Society for Information Science and Technology* 36:30–3.
Munck, Renaldo. 2013. 'The precariat: a view from the South'. *Third World Quarterly* 34 (5):747–62.
Naughton, John. 2014, 20 April. 'Why Facebook and Google are buying into drones'. *Guardian*, www.theguardian.com/world/2014/apr/20/facebook-google-buying-into-drones-profit-motive.
Negroponte, Nicholas. 2003 [1975]. 'From "soft architecture machines"'. In *theNEWMEDIAREADER*, ed. Noah Wardrip-Fruin and Nick Montfort, 353–65. Cambridge, MA & London: The MIT Press.
Negroponte, Nicholas. 2006, February. 'One laptop per child' [video]. TED, www.ted.com/talks/nicholas_negroponte_on_one_laptop_per_child.
Negroponte, Nicholas. 2007. 'One laptop per child, two years on' [video]. TED, www.ted.com/talks/nicholas_negroponte_on_one_laptop_per_child_two_years_on#t-13632.
Neocleous, Mark. 2013. 'Police power, all the way to heaven: *cujus est solum* and the no-fly zone'. *Radical Philosophy* (182):5–14.
Nielsen. 29 January 2015. 'Moving to mobile: how phones are revolutionizing hunger relief'. Nielsen, www.nielsen.com/us/en/insights/news/2015/moving-to-mobile-how-phones-are-revolutionizing-hunger-relief.html.

NLR. 1968. 'Introduction to special issue on France May 1968'. *New Left Review* (52):1–8.
Nordstrom, Carolyn. 2000. 'Shadows and sovereigns'. *Theory, Culture and Society* 17 (4):35–54.
Noys, Benjamin. 2012. *The Persistence of the Negative: A Critique of Contemporary Continental Theory*. University of Edinburgh Press.
Nussbaum, Bruce. 6 July 2010. 'Is humanitarian design the new imperialism?' Fast Company, www.fastcodesign.com/1661859/is-humanitarian-design-the-new-imperialism.
O'Brien, Jay. 1983. 'The formation of the agricultural labour force in Sudan'. *Review of African Political Economy* 10 (26):15–34.
O'Brien, Jay. 1985. 'Sowing the seeds of famine: the political economy of food deficits in Sudan'. *Review of African Political Economy* (33):23–32.
O'Callaghan, Derek, Nico Prucha, Derek Greene, Maura Conway, Joe Carthy and Padraig Cunningham. 2014, 29 January. 'Online social media in the Syria conflict: encompassing the extremes and in-between', http://arxiv.org/pdf/1401.7535v1.pdf.
ODI. 2015. 'Doing cash differently: how cash transfers can transform humanitarian aid'. Report of the High Level Panel on Humanitarian Cash Transfers. London: Overseas Development Institute.
OECD. 2008. *Growing Unequal? Income Distribution and Poverty in OECD Countries*. Paris: Organisation of Economic Cooperation and Development.
OECD. 2011. 'Future global shocks: improving risk governance'. Organisation of Economic Cooperation and Development, www.oecd.org/governance/48329024.pdf.
O'Malley, Pat. 2009. '"Uncertainty makes us free": liberalism, risk and individual security'. *Behemoth. A Journal on Civilisation* 2 (3):24–38.
O'Malley, Pat. 2010. 'Resilient subjects: uncertainty, warfare and liberalism'. *Economy and Society* 39 (4):488–509.
O'Reilly, Tim. 2005, 30 September. 'What is Web 2.0: design patterns and business models for the next generation of software', www.oreilly.com/pub/a/web2/archive/what-is-web-20.html.
Oxfam. 2016. *An Economy for the 1%: How Privilege and Power in the Economy Drive Extreme Inequality and How This Can Be Stopped*. Oxford: Oxfam International.
Packard, Vance. 1957. *The Hidden Persuaders*. New York: Random House.
Page, Tim. 1989. *Page after Page: Memoirs of a War-torn Photographer*. London: Atheneum.
Palen, Leysia, Kenneth Anderson, Gloria Mark, et al. 2010. 'A vision for technology-mediated support for public participation & assistance in mass emergencies & disasters'. In *Proceedings of ACM-BCS Visions of Computer Science 2010*, 1–12. British Informatics Society, https://dl.acm.org/citation.cfm?id=1811194.
Parks, Lisa. 2009. 'Digging into Google Earth: an analysis of "Crisis Darfur"'. *Geoforum* 40:535–45.
Pelosky, Robert J. 2002. 'Synchronous failure: the real danger of the 21st century'. Paper presented at The Elliot School of International Affairs, George

Washington University, Washington DC, Distinguished Speakers Series, 4 December, https://dl.acm.org/citation.cfm?id=1811194.
Petti, Alessandro. 2007. *Arcipelaghi e enclave: architettura dell'ordinamento spaziale, contemporaneo*. Milan: Bruno Mondadori.
Petti, Alesandro. 2008. 'Asymmetries in globalised space'. Paper presented at the International Workshop on Power, Governmentality, Resistance and State of Exception in the Arab World, Beirut, August.
Piketty, Thomas. 2014. *Capital in the Twenty-First Century*. Cambridge, MA: Harvard University Press.
Pirlot, Alexandrine. 2014, 21 January. 'Big data: a tool for development or threat to privacy?' Privacy International, https://www.privacyinternational.org/blog/big-data-a-tool-for-development-or-threat-to-privacy.
Postlewait, Heidi, Kenneth Cain and Andrew Thompson. 2006. *Emergency Sex and Other Desperate Measures*. London: Ebury Press.
Prahalad, C. K. 2006. *The Fortune at the Bottom of the Pyramid: Eradicating Poverty Through Profits*. Upper Saddle River, NJ: Wharton School of Publishing
Prins, E. 2008. 'Use of low cost Landsat ETM+ to spot burnt villages in Darfur, Sudan'. *International Journal of Remote Sensing* 28 (4):1207–14.
PRIO. 2009. 'Data on armed conflict'. Peace Research Institute Oslo (PRIO), https://www.prio.org/Data/Armed-Conflict.
Pupavac, Vanessa. 2001. 'Therapeutic governance: psycho-social intervention and trauma risk management'. *Disasters* 25 (4):358–72.
Pupavac, Vanessa. 2012. *Language Rights: From Free Speech to Linguistic Governance*. Basingstoke: Macmillan/Palgrave.
Rabinow, Paul. 1995. *French Modern: Norms and Form of the Social Environment*. University of Chicago Press.
Radio Dabanga. 2009. 'On the ground reporter' [online computer game]. Radio Dabanga with Butch and Sundance Media, Directionaldesign.nl, https://www.radiodabanga.org/darfurgame/english/game.html#.
Rahebi, Mohammad-Ali. 2015. 'Biopolitical immanence or whether Foucault and Deleuze still matter'. *La Deleuziana – Online Journal of Philosophy (Crisis of the European Biopolitics)* (1): 73–90.
Ramalingam, Ben. 2014, 5 December. 'The WDR 2015: putting development on the couch?' Aid on the Edge of Chaos: Rethinking International Cooperation in a Complex World, https://aidontheedge.info/2014/12/05/the-wdr-2015-putting-development-on-the-couch.
Ramalingam, Ben, Miguel Laric and John Primrose. 2014. 'From best practice to best fit: understanding and navigating wicked problems in international development'. Working Paper. London: Overseas Development Institute.
Read, Jason. 2016. *The Politics of Transindividuality*. Chicago, IL: Haymarket Books.
Redfield, Peter. 2015. 'Fluid technologies: the Bush Pump, the Life Straw® and microworlds of humanitarian design'. *Social Studies of Science* 46 (2):159–83, http://journals.sagepub.com/doi/pdf/10.1177/0306312715620061.
RedR. 2017. 'Sudan training programme'. Register of Engineers for Disaster Relief (RedR), www.redr.org.uk/en/wherewework/sudan.cfm.
Reid, Julian. 2006. *The Biopolitics of the War on Terror: Life Stuggles, Liberal*

Modernity and the Defence of Logistical Societies (Reappraising the Political). Manchester University Press.
Reid, Julian. 2009. 'Politicizing connectivity: beyond the biopolitics of information technology in international relations'. *Cambridge Review of International Affairs* 22 (4):607–23.
Richmond, Oliver. 2014. *Failed Statebuilding: Intervention, the State, and the Dynamics of Peace Formation*. New Haven, CT, & London: Yale University Press.
ROAPE. 1974. 'Editorial'. *Review of African Political Economy* (1):1–8.
Robins, Steven. 2014. 'Poo wars as matter out of place: "Toilets for Africa" in Cape Town'. *Anthropology Today* 30 (1):1–3.
Rogers, Colin. 2006. 'Accessing the inaccessible: the use of remote programming strategies in highly insecure countries to ensure the provision of humanitarian assistance. Iraq: a case study'. Dissertation for MA in Post-war Recovery Studies, Post-war Reconstruction and Development Unit (PRDU), University of York.
Roitman, Janet, ed. 2001. 'New sovereigns? The frontiers of wealth creation and regulatory authority in the Chad Basin'. In *Intervention and Transnationalism in Africa: Global–Local Networks of Power*, ed. Thomas Callaghy, Ronald Kassimir and Robert Letham, 190–215. Cambridge University Press.
Rose, Nikolas. 1996. 'The death of the social? Re-figuring the territory of government'. *Economy and Society* 25 (3):327–56.
Rose, Nikolas. 2000. *Powers of Freedom: Reframing Political Thought*. Cambridge University Press.
Roser, Max. 2015. 'Income inequality'. OurWorldInData.org, http://ourworldindata.org/data/growth-and-distribution-of-prosperity/income-inequality.
Rostow, W. W. 1960. *The Stages of Economic Growth: A Non-Communist Manifesto*. Cambridge University Press.
Rouvroy, Antoinette. 2009. 'Governmentality in an age of automatic computing: technology, virtuality and Utopia'. SelectedWorks, https://works.bepress.com/antoinette_rouvroy/26.
Rouvroy, Antoinette. 2012. 'The end(s) of critique: data-behaviourism vs. due-process'. In *Privacy, Due Process and the Computational Turn*, ed. Mirelle Hildebrandt and Ekatarina De Vries, 1–19. London: Routledge.
Ruel, Marie T., and Harold Alderman. 2013. 'Maternal and child nutrition 3: nutrition-sensitive interventions and programmes. How can they help to accelerate progress in improving maternal and child nutrition?' *Lancet* (382):536–51.
Sabaratnam, Meera. 2013. 'Avatars of Eurocentrism in the critique of liberal peace'. *Security Dialogue* 44 (3):259–78.
Schumacher, E F. 1974. *Small is Beautiful: A Study of Economics as if People Mattered*. London: Abacus.
Schwittay, Anke. 2011. 'The marketization of poverty'. *Current Anthropology* 52 (3).
Scott-Smith, Tom. 2013. 'The fetishism of humanitarian objects and the management of malnutrition in emergencies.' *Third World Quarterly* 34 (5):913–28.
Scott-Smith, Tom. 2016. 'Humanitarian neophilia: the "innovation turn" and its implications'. *Third World Quarterly* 37 (12):2229–51.

Sen, A. K. 1981. *Poverty and Famines*. Oxford: Clarendon.
Setel, Philip W., Sarah B. Macfarlane, Simon Szreter, et al. 2007. 'A scandal of invisibility: making everyone count by counting everyone'. *Lancet* 370 (3):1569–77.
Simondon, Gilbert. 2009. 'Technical mentality'. *Parrhesia: A Journal of Critical Philosophy* 7:17–27.
Skinner, Danielle. 2010, 14 December. 'Dr Steven Livingston speaks on emerging communication technology in Africa'. US Africa Command: US Africom Public Affairs, www.africom.mil/printStory.asp?art=5761.
Slim, Hugo. 2015, 15 March. 'Eye scan therefore I am: the individualization of humanitarian aid'. European University Institute, http://iow.eui.eu/2015/03/eye-scan-therefore-i-am-the-individualization-of-humanitarian-aid.
Sloterdijk, Peter. 2013 [2005]. *In the World Interior of Capital*, trans. Wieland Hoban. Cambridge & Malden: Polity.
Smirl, Lisa. 2008. 'Building the other, constructing ourselves: spatial dimensions of international humanitarian response'. *International Politics Sociology* 2:236–53.
Smith, Jason E. April 2017. 'Nowhere to go: automation, then and now part two'. The Brooklyn Rail, http://brooklynrail.org/2017/04/field-notes/Nowhere-to-Go-Automation-Then-and-Now-Part-Two.
Song, Steve. April 2015. 'African undersea cables'. Village Teleco, https://manypossibilities.net/african-undersea-cables.
Sorensen, Jens Stilhoff. 2014. 'The return of plural society: statebuilding and social splintering'. *Journal of Peacebuilding* 2 (3):270–85.
Spencer, Douglas. 2016a. *The Architecture of Neoliberalism: How Contemporary Architecture Became an Instrument of Control and Compliance*. London & Oxford: Bloosmbury.
Spencer, Douglas. 2016b, 29 September. 'Out of the loop'. Spatial Register, https://spatialregister.wordpress.com/2016/09/29/out-of-the-loop-new-essay-in-volume-49.
Spiz, Pierre. 1978. 'Silent violence: famine and inequality'. *International Social Science Journal* 30 (4):867–92.
Spreeuwenberg, Kimberley, and Thomas Poell. 2012, June. 'Android and the political economy of the mobile internet: a renewal of open source critique'. First Monday, http://firstmonday.org/ojs/index.php/fm/article/view/4050/3271#author.
Srnicek, Nick. 2016. *Platform Capitalism*. Cambridge: Polity.
SSP. 2012, 24 July. 'Monitoring the crisis in the Sudans'. Satellite Sentinel Project: Enough, Washington DC, http://satsentinel.org.
Stadler, Max. 2014. 'Neurohistory is bunk? The not-so-deep history of the postclassical mind'. *Isis: An International Review Devoted to the History of Science and Its Cultural Influences* 105 (1):133–44.
Stiegler, Bernard. 2010. *For a New Critique of Political Economy*. Cambridge: Polity.
Stiegler, Bernard. 2014 [2006]. *The Lost Spirit of Capitalism*, volume III, trans. Daniel Ross. Cambridge: Polity.
Stiegler, Bernard. 2016. *Automatic Society*, volume 1: *The Future of Work*. Cambridge: Polity.

STJ. 2000. *The Black Book: Imbalance of Power and Wealth in Sudan.* Khartoum: Seekers of Truth and Justice.
Stoddard, Abby, Adele Harmer and Victoria DiDomenico. 2009. *Providing Aid in Insecure Environments: 2009 Update.* London: Humanitarian Policy Group (HPG), Overseas Development Institute.
Stoddard, Abby, Adele Harmer and Katherine Haver. 2006. *Providing Aid in Insecure Environments: Trends in Policy and Operations.* London: Humanitarian Policy Group, Overseas Development Institute.
Stoddard, Abby, Adele Harmer and Jean S. Renouf. 2010. *Once Removed: Lessons and Challenges in Remote Management of Humanitarian Operations for Insecure Areas.* London and New York: Humanitarian Outcomes.
Stoddard, Abby, Adele Harmer and Kathleen Ryou. 2014. 'Aid worker security report 2014, unsafe passage: road attacks and their impact on humanitarian operations'. London: Humanitarian Outcomes.
Stover, Eric. 1987. 'A famine of wandering'. *New Scientist* (1589):30–1.
Streeck, Wolfgang. 2011. 'The crises of democratic capitalism'. *New Left Review* (71):5–29.
Sulik, John J., and Scott Edwards. 2010. 'Feature extraction for Darfur: geospatial applications in the documentation of human rights abuse'. *International Journal of Remote Sensing* 31 (10):2521–33.
Summerfield, Derek. 1996. 'The impact of war and atrocity on civilian populations: basic principles for NGO interventions and a critique of psychosocial trauma projects'. Network Paper No. 14. London: Overseas Development Institute, Relief & Rehabilitation Network.
Taplin, Jonathan. 2017. *Move Fast and Break Things: How Facebook, Google and Amazon have Cornered Culture and What it Means for All of Us.* London: Macmillan.
Taylor, Linnet. 2013, 25 October. 'Surveil the rich, observe the poor: big data at the Internet Governance Forum 2013'. linnettaylor.wordpress.com, http://linnettaylor.wordpress.com/2013/10/25/surveil-the-rich-observe-the-poor-big-data-at-the-internet-governance-forum-2013.
Terranova, Tiziana. 2004. *Network Culture: Politics for the Information Age.* Ann Arbor, MI: Pluto Press.
Thaler, Rodger, and Cass Sunstein. 2008. *Nudge.* London: Penguin.
Tofler, Alvin. 1980. *Future Shock: The Third Wave* New York: Bantam Books.
TUC. 2017. *The Impact of Increased Self-Employment and Insecure Work on the Public Finances.* London: Trade Union Congress.
Turner, Fred. 2006. *From Counterculture to Cyberculture: Stewart Brand, the Whole Earth Network, and the Rise of Digital Utopianism.* Chicago & London: University of Chicago Press.
Tvedt, Terje. 1994. 'The collapse of the state in Southern Sudan after the Addis Ababa Agreement: a study of the internal causes and role of NGOs'. In *Short-Cut to Decay: The Case of Sudan,* ed. Sharif Harir and Terje Tvedt, 68–103. Upsalla: Nordiska Afrikainstitutet.
UN. 1994. *Strengthening of the Coordination of Humanitarian Emergency Assistance of the United Nations.* New York: United Nations.
UN. 2001. *Safety and Security of Humanitarian Personnel and Protection of United Nations Personnel.* New York: United Nations General Assembly.

UN. 2004. 'Living with risk: a global review of disaster reduction methods'. International Strategy for Disaster Reduction (ISDR). New York: United Nations.
UN. 2011. 'The global social crisis: report on the world social situation 2011'. New York: Department of Economic and Social Affairs, United Nations.
UNASF. 2006. 'Advanced security in the field' [CD-ROM]. United Nations.
UNBSF. 2003. 'Basic security in the field' [CD-ROM]. United Nations.
UNDP. 1994. 'Position paper of the Working Group on Operational Aspects of the Relief to Development Continuum'. New York: UNDP.
UNDP. 2008. *Creating Value for All: Strategies for Doing Business with the Poor*. New York: United Nations Development Programme
UNECA. 2015a. *Country Profile 2015: Kenya*. Addis Ababa: United Nations Economic Commission for Africa.
UNECA. 2015b. 'Economic report on Africa 2015: industrialising through trade'. Addis Ababa: United Nations Economic Commission for Africa.
UNGP. 2009. 'United Nations Global Pulse: harnessing innovation to protect the vulnerable'. United Nations Global Pulse, www.unglobalpulse.org.
UN-Habitat. 2003. 'The challenge of slums: global report on human settlements 2003'. London & Sterling, VA: Earthscan for UN-Habitat.
UN-Habitat. 2016. 'Slum almanac: 2015–2016'. Participatory Slum Upgrading Programme. Nairobi: UN-Habitat.
UNHCR. 1968. 'Agreement between the United Nations High Commissioner for Refugees and the Government of the Republic of Sudan concerning the establishment of a branch office of the High Commissioner in Khartoum', Registry Fonds 11, Series 1, Code 4/9 SUD. Geneva: United Nations High Commissioner for Refugees.
UNHCR. 2008. 'Google Earth layers'. UN High Commissioner for Refugees, www.unhcr.org/pages/49c3646c4d3.html.
UNHCR. 2017a. 'Figures at a glance'. UNHCR, www.unhcr.org/uk/figures-at-a-glance.html.
UNHCR. 2017b. 'Searching for Syria' [interactive video]. UNHCR in partnership with Google, https://searchingforsyria.org/en/what-was-syria-like-before-the-war.
UNOCHA. 2013. *Humanitarianism in the Network Age*. OCHA Policy and Study Series. New York: UN Office for Coordination of Humanitarian Affairs.
Urry, John. 2003. *Global Complexity*. Cambridge: Polity.
USAID. 2013, 6 June. 'USAID and DFID announce global development innovation ventures to invest in breakthrough solutions to world poverty'. https://www.usaid.gov/news-information/press-releases/usaid-and-dfid-announce-global-development-innovation-ventures.
Usborne, David. 2004, 18 January. 'Coalition uses 1918 British report on tribal systems'. *Independent*, www.independent.co.uk/news/world/americas/coalition-uses-1918-british-report-on-tribal-system-74180.html.
Van Brabant, Koenraad. 1998. 'Cool ground for aid providers: towards better security management in aid agencies'. *Disasters* 22 (2):109–25.
Van Brabant, Koenraad. 1999. 'Security training: where are we now?' *Forced Migration Review* (4):7–10.
Van Brabant, Koenraad. 2000. *Operational Security Management in Violent*

Environments: A Field Manual for Aid Agencies. London: Humanitarian Policy Network (HPN), Overseas Development Institute.

Verjee, Firoz. 2005. *The Application of Geomatics in Complex Humanitarian Emergencies*. Washington, DC: The George Washington University, Institute for Crisis, Disaster & Risk Management.

Verjee, Firoz. 2007. 'An assessment of the utility of GIS-based analysis to support the coordination of humanitarian assistance'. Dissertation submitted in partial satisfaction of the requirements for the degree of Doctor of Science, The School of Engineering and Applied Science, The George Washington University, Washington, DC.

Virilio, Paul. 2007 [1977]. *Speed and Politics: An Essay on Dromology*. Los Angeles: Semiotext(e).

Walker, Peter. 1987. *Food for Recovery: Food Monitoring and Targeting in Red Sea Province, Sudan, 1985–1986*. Oxford: Oxfam.

Walker, Peter. 2009 [1989]. *Famine Early Warning Systems: Victims and Destitution*. Abingdon & New York: Earthscan. Original edition, 1989.

Walton, John, and David Seddon. 1994. *Free Markets and Food Riots: The Politics of Global Adjustment*. Oxford: Blackwell.

Washington Times. 2009, 16 June. 'Iran's Twitter revolution'. www.washingtontimes.com/news/2009/jun/16/irans-twitter-revolution.

Weber, Max. 2005 [1930]. *The Protestant Ethic and the Spirit of Capitalism*, trans. Talcott Parsons. London & New York: Routledge Classics.

WEF. 2010. 'Global risks 2010: a global risk network report'. Geneva: World Economic Forum.

WEF. 2016, 19 October. 'These are the world's five biggest slums'. World Economic Forum, www.weforum.org/agenda/2016/10/these-are-the-worlds-five-biggest-slums.

Weizman, Eyal. 2002, 24 April. 'The politics of verticality'. openDemocracy, https://www.opendemocracy.net/ecology-politicsverticality/article_801.jsp.

Whitaker, Ben. 1983. *A Bridge of People: A Personal View of Oxfam's First Forty Years*. London: Heinemann.

Wiener, Norbert. 1954. *The Human Use of Human Beings: Cybernetics and Society*. Boston: Houghton Mifflin Co.

Wiley, David. 2012. 'Militarizing Africa and African Studies and the US Africanist response'. *African Studies Review* 55 (2):147–61.

Winer, Nicholas. 2008. *The Tethered Goat*. YouWriteOn.com, https://www.amazon.com/Tethered-Goat-Nicholas-Winer-ebook/dp/B005H3EWKY.

Wisner, Ben, Piers Blaikie, Terry Cannon and Ian Davis. 2004. *At Risk: Natural Hazards, People's Vulnerability and Disasters*. London & New York: Routledge.

Wolpe, Harold. 1972. 'Capitalism and cheap labour power in South Africa: From Segregation to Apartheid'. *Economy and Society* 1:425–56.

Wood, Denis. 2010. *Rethinking the Power of Maps*. New York: The Guilford Press.

Woodis, Jack. 1960. *Africa: The Roots of Revolt*. London: Lawrence and Wishart.

World Bank. 2015. *World Development Report 2015: Mind, Society, and Behaviour*. Washington, DC: World Bank.

World Bank. 2016a. *Cash Transfers in Humanitarian Contests*. Washington DC: The World Bank.
World Bank. 2016b. *Taking on Inequality: Poverty and Shared Prosperity 2016*. Washington, DC: World Bank.
World Bank. 2017. *The State of Identification Systems in Africa: A Synthesis of Country Assessments*. Washington, DC: World Bank.
Zuckerberg, Mark. 2014, 28 March. 'Connecting the world from the sky'. FaceBook, http://internet.org/press/connecting-the-world-from-the-sky.
Zyck, Steven A., and Randolph Kent. 2014. 'Humanitarian crises, emergency preparedness and response: the role of business and the private sector – Final report'. London: Humanitarian Policy Group (HPG), Overseas Development Insitute (ODI).

INDEX

'A Cautious Prometheus? A Few Steps Toward a Philosophy of Design' (Latour) 187–8
Abu Manga, Al-Amin 55
ACAS *see* Association of Concerned Africa Scholars
Achilles Initiative 110
Ackerman, R. K. 145
Ackerman, S. 102
Adorno, T. 34
Afghanistan 13, 100, 145, 196n4, 196n5
Africa 31, 45, 120, 125, 134–5, 164
African Studies Association (ASA) 45
Agamben, G. 64, 65, 68, 69, 132, 167, 195n7
Age of Anger (Mishra) 190
Agencies Mississippi 112
Agier, M. 63
Ahmed, T. M. 70
AI *see* Amnesty International
Airbnb 201n8
Alcock, R. 21, 179, 183
Alderman, H. 172, 173
Algeria 2, 30
Althusser, L. 51
Amazon 201n8
American Anthropological Association 196n4
Amnesty International (AI) 146, 149, 201n10
Amoore, L. 9, 66
Amsden, A. H. 15, 137

Anderson, C. 9
Andreotti, L. 66
Angola 2, 30
Anthropocene era 74, 174, 188
anthropology
  as individual sink-or-swim exercise 46–9
  new practices in 43–6, 196n4
  structuralist method 49–53
Aquila drone technology 136
Architecture Machine Group (later the Media Lab) 90, 199n8
Arendt, H. 7–8, 11, 190–1
Argentina 42
ASA *see* African Studies Association
Asad, T. 44
Associated Press 136
Association of Concerned Africa Scholars (ACAS) 45
Australia 6, 31
automatic society 129
Ayers, A. 26, 114–15, 123, 137, 170, 181

Baghdad 103–4
Balakrishnan, G. 131, 132
Balibar, E. 51, 69, 123
Bally, P. J. B. 146, 151
Banaji, J. 44, 46
Band Aid 59–60, 61
Bannaga, S. E. I. 54
Barbrook, R. 20, 141
Barnes, C. 86

INDEX

Barnett, A. 42, 61
Bartlett, J. 140
Batty, D. 148
Beck, U. 26
Becker, K. F. 127
behavioural economics 183, 186
Beijing Review 137
Bell, D. 187
Berlin 42
Better Shelter 162
Betts, A. 116, 153, 159, 160
Big Data 70, 123, 128, 130, 151, 152, 182, 189
biohuman 14, 22, 26, 51, 69, 110, 129, 147, 151, 176
biometric technologies 64, 162, 164, 166, 167
biopolitics 22, 66, 69, 110, 115, 129, 165, 172, 176, 194n6
Bitcoin 164
Bjorgo, E. 148, 150
Black, R. E. 173
Black Book: Imbalance of Power and Wealth in Sudan (STJ) 198n13
Blackburn, R. 43
Blanchetiere, P. 109
Bloom, L. 116, 153, 160
Boltanski, L. 17, 18, 19, 20, 23, 24, 25, 58–9, 79–80, 88, 123, 184
Bonneuil, C. 188
boomerang effect 10–11, 13, 27, 115, 178, 193n6
Booth, D. 71, 72
Bosnia 102, 199n3
Bostdorff, D. M. 31
Bottom of the Pyramid (BOP) economics 125, 126, 140, 160
Bouchardy, J.-Y. 145, 148, 150
Boutang, Y. M. 8, 184
Boutros-Ghali, B. 100
Bradbury, M. 97
Braidotti, R. 10, 105, 130
Brazil 2, 30
Brexit 7
Briend, A. 171
Brown, W. 6, 140
Bruderlein, C. 99, 101, 104
Bryant, R. L. 86

Brynjolfsson, E. 39
Buchanan, K. 2, 30
Buck-Morss, S. 34
Burns, R. 156
Bush, R. 31, 140
bush air strips 97, 200n4
Butler, G. 107

CableFree 136
Cadwalladr, C. 7, 184
Cambodia 59
Cameron, A. 20
Canada 6, 31
capitalism
  boot strap 124–5
  and computational networks 69–70
  decline in academic interest in 17
  disappearance of 71
  inclusive 125
  Marcusian disappearance of 33–4
  modern iterations of 18, 194n3 & n4
  opposition to 41
  post-social 137
  re-engagement with 17–18
  and shift towards medicalization 174
  spirit of 17–26, 79
caretaker society 186–9
Caribbean 31
Carr, N. 178
Castells, M. 15, 79
catastrophism 3, 36
Cater, N. 72–3, 76
Cederstrom, C. 23
Chad 146, 148
Chamayou, G. 102, 113, 145, 189
Chan, J. 153
Chandler, D. 3, 7, 9, 10, 74, 75, 104, 130, 174
Chiapello, E. 17, 18, 19, 20, 23, 24, 25, 58–9, 79–80, 88, 123
China 2, 30
Chouliaraki, L. 3, 32, 174
circulation
  barriers to / closure of 6–7, 88, 93
  and computational turn 11
  entropy of 36

227

circulation (*cont.*)
  as factor of spatial organization 3–4
  and fieldwork in Sudan 46–55
  Foucauldian 4
  freedom/openness to 31, 38, 46, 59, 93
  and ground friction 1933
  maintenance of 131
  mobility vs immobility 24–6
  and New Left's view of revolution 29
  open to accidents, dangers, unforeseen circumstances 5
  reduction in 96–9
  as robotic 4–5
  and security 4, 5, 66–9
  shift to connectivity 5–6, 13, 130
Clark, T. 121
class 20, 24–5, 28, 29, 30, 39, 40, 41, 45–6, 48, 59, 79, 122, 123, 184, 194n3 & n4
climate change 7
Clooney, George 149
cognitive development 182–4, 185–6
cognitive turn 179–80
  feedback 182–4
  and mental precarity 180–2
  optimizing reproduction 184–6
Cold War 6, 18, 31, 59, 60, 93, 100, 188, 196n5, 200n1
Collier, S. J. 144
Collins, P. 116
Collins, S. 171
Collinson, S. 32, 61, 98, 100, 147
Committee of Concerned Asia Scholars 45
communities
  action-orientation 83–4, 168
  and constructivist learning 90–2
  continuous enrolment in prototyping 171–3
  as creative enterprise 83–4
  and digital decentralizing power 156–7
  and environmental uncertainty 86–7
  exploitation/mobility differential 87–8
  and humanitarian design/innovation 169–70
  and local knowledge 88–90
  made visible by NGOs 85–6
  mobility differential 87–8
  and ownership of projects 84, 169
  preparing/empowering 169
  and privileging of marginalized/excluded 86–7
  projects as key form of intervention in 85–7
  as self-acting 84
  self-reliance 86
  and smart technologies 170–1
Comoretto, A. 109
complex emergency 72–4, 198n16
complexity 71–2
  early warning 74–6
  environmental exposure 72–4
  and undermining of relief/development 84–5
computational turn 10–11, 14, 62, 83, 113, 130, 177
computer learning 90–2
computers
  diffusion of 15, 152
  as provider/guarantor of freedom, choice, personal creativity 152
Conneally, P. 153
connectivity
  in Africa 134–5, 138–40
  and circulation 3–6, 24
  data-based 4–5
  electronic atmosphere 133–6
  expanding the enclave 136–40
  global 156
  increase in 104, 111
  as infrastructural/ontological 143–4
  in Kenya 138–40
  and move beyond project form 127
  need for speed 24
  paradox of 189–92
  and remoteness 6–8
  and sense of democratization/empowerment 133
  as shared 32
  spread of 132, 133–6, 143
Coole, D. 130

INDEX

Cooper, M. 21, 36, 37
Corlett, A. 26, 121
Cornia, G. A. 117
Cortada, J. W. 6, 15, 16
counter-culture 12, 28, 40, 91, 151
Coward, M. 120
Cowen, D. 16, 137
Crary, J. 24, 116
Crisis in Darfur initiative 148–9
crisis informatics 158, 165
  academic exploration or potential 153
  background 151–3
  decentralizing power 156–7
  development/assumptions 153–4
  as distinct area of operational possibility 152
  distributed information systems 156
  enabling resilience 154–5
  growing impact 154
  and *Homo inscius* 155
  interconnection with cognitive development 182–3
  provides means/rationale for information feedback 159
Critical Asian Studies 45
Cross, J. 160, 162, 175
Crowe, A. 163
Crowley, J. 153
Crutcher, M. 152
Cuba 2, 30, 40
Cultivation of Hunger, The (Ahmed) 70
Cuppens, Y. 87
Cutler, P. 75, 77
Cutts, M. 102
cybernetic *episteme* 5, 21, 27, 29, 58, 67, 72, 168
  approach to entropy 36–7
  early warning as central to 76–7
  etymological origins 68
  expansion of 144
  and exploitation 87
  first-order/second-order 36–7, 195n8
  origins/development 35–6
  and the project form 80

  shaping of 33
  singularity of vision 176–7
  steersman analogy 68–9
  as universal discipline 37
  universities as contested space 40–3
  universities as instrumental in rise/dissemination of 39, 105n1

da Costa, D. F. 98
Daly, M. W. 96
Darfur 75, 146–50, 196n5
data 8–10, 51–2, 76, 123, 151, 156, 191
data behaviourism 9
data informatics 116
Davis, M. 31, 116, 124–5, 133
de Bruijin, M. 134
Dean, M. 23
decolonization 2, 6, 29–32, 44
deindustralization 15, 136–7, 196n2
Deleuze, G. 66
Department of Peace-keeping Operations (DPKO) 103, 104
Development 2.0 134
Development Initiatives 120
developmentalism 62
  as automatic vacuum cleaner 89
  as combination of paternalism / techno-pastoral aesthetic 128
  definition of 89–90
  and neoliberalism 126
  and the poor 81–2
  spatial psychosis concerning 127–8
DFID 111, 116, 124, 171
Dharavi slum (Mumbai) 125
digital streamlining 14, 23, 123, 140, 167–8, 179, 183, 186
DigitalGlobe 149
Dillon, M. 37, 51, 129
Dingle, A. 102
disaster/s 2–3
  and crisis informatics 154–5
  and humanitarian technologies 166
  management 77
  modernist views of 62–3
  new ontology of 62–70
distributed information systems 156, 158

229

'Do They Know It's Christmas?' 59
Dobbs, L. 149
Doing Development Differently 89, 179, 199n6
Donini, A. 101
Donnell, H. 24, 140
Donovan, K. P. 134, 162, 167
DPKO *see* Department of Peacekeeping Operations
drones 136, 163
Duffield, M. 19, 30, 32, 46, 49, 53–4, 59, 61, 72, 85, 86, 87, 94, 96, 98, 100, 113, 117, 124, 167
Dupuy, J.-P. 21

Eagle, Pentland, 152
early warning systems 74–7, 198n17 & n18
Easterling, K. 132, 133, 134–5, 136, 137, 138, 139–40, 141
EC *see* European Community
ECHO 107
Edkins, J. 75
Edwards, M. 71, 74, 89–90
Edwards, P. N. 7, 188
Edwards, S. 146, 147, 149
Egypt 30
Eide, E. B. 100
Elden, S. 24, 113
Eldredge, E. 77
Elhaway, S. 147
'The end of history' (Fukuyama) 186–7
The End of Ideology: On the Exhaustion of Political ideas in the Fifties (Bell) 187
Enough Project 149
entropy resistance 35–8
ENVIREF *see* Environmental Monitoring of Refugee Camps
Environmental Monitoring of Refugee Camps (ENVIREF) 146
equilibrium theory 194n6
Eritrea 61
Errington, F. 163
Escudero, Ruben Salgado 142
Esquerre, A. 184
ethical irony 174–6

Ethiopia 61
EU Joint Research Centre (JRC) 145
European Community (EC) 85
Evans, B. 22, 124
exploitation 24–5, 194n7
Eyes on Darfur initiative 149

Facebook 135, 136, 156, 201n8
fake news 175, 183
famine 59–60, 67–8, 70, 173
  as complex emergency 72–4
  early warning of 74–6, 198n17 & n18
  early warning systems 76–7, 198n16
  Sahel 75
fantastic invasion 78
  action-orientation as project/community 83
  and direct humanitarian action 58–62, 98, 168
  and expanding cybernetic *episteme* 144
  ideological completeness of 67, 197n10
  key assumptions concerning crisis informatics 153–4, 157
  legitimacy of 71
  and new ontology of disaster 62–70
  origin of term 196n1
  and peasant dispossession/impoverishment 123–4
  and progressive neoliberalism 126
  and projects as means of information transference 83–4, 159–60
  and second wave of terrestrial globalization 93
  and shift from theory to complexity-thinking 70–7, 173
  in Sudan 57–8
  and transformation of knowledge 78
  work progressed by humanitarian innovation 167
FAO *see* United Nations Food and Agriculture Organization
Faris, J. 45

financial crisis (2008-9) 119–20, 137, 180
Fleming, P. 23
Fleming, S. 86
Focus on Technology and the Future of Humanitarian Intervention Report (IFRC 2013) 158
Foer, F. 90, 178, 182
food aid 172–4
Fordism 18, 23, 54, 90, 194n5, 196n2
fortified aid compounds 112–13, 147, 156
  defensive architectural style 95–6
  description of 94–5
  environmental impact 95
  reasons for 99–102
  and reduced circulation 96–9
  and security systems / training programmes 102–4
  spread of 96, 199–200n3
  withdrawal into 94
Fortune at the Bottom of the Pyramid: Eradicating Poverty Through Profits, The (Prahalad) 125
Foucault, M. 4, 11, 22, 33, 49–50, 58, 66, 67, 68, 69
Fox, F. 100
Frankfurt School 34
Franklin, V. 18
Fraser, N. 20, 85, 116, 126
Fredriksen, A. 162
free space optics communications (FSO) 136
Freedom-As-A-Service 164
Fressoz, J.-B. 188
Friedman, E. D. 119
FSO *see* free space optics communications
Fukuyama, F. 34, 66, 186, 187
Future Shock (Tofler) 141

Gabrys, J. 143, 170
Galloway, A. R. 9, 10, 33, 130
Gassmann, P. 99, 101, 104
Geldof, B. 60, 165
GeoEye 149

geospatial technology 145–7
Geotz, T. 183
Gibson, W. 129
gig economy 127–8
Gillula, J. 163
Giroux, J. 155
Global Awareness 152
global North 80
  casualization in 115
  death of the social in 26, 65
  emergence of enrichment economies in 184
  job(less) concerns in 121–3
  North–South interface 2–3, 11, 16, 23, 27, 32
  personalized consumption in 16, 83
  post-social work/career structures 168, 184
  role of disasters in 2–3
  secondary market for humanitarian objects in 175–6
  and smart technology 133
  social, urban, political reform in 22
  spread of computers in 15
  and use of crisis informatics 152–3
  use of term 193n2
Global Positioning System (GPS) 144, 200n1, 201n2 & n3
global precariat 13–14, 74, 88
  associated with rapid post-colonial growth of slums 116–17
  and automation 122–3
  and blurring of North–South distinctions 121–3, 190
  and boot strap capitalism 124–5
  definition of precarity 115–16
  and fantastic invasion 126–7
  and future of network capitalism 140
  as historically novel form of dispensation 129
  Kenyan 138–9
  lack of civil registration among 164
  and mobile connectivity 130–1
  and recycling poverty 123–6
  and registers of informality 117–21
  and shadow economies 124
  and social reproduction 116

231

global precariat (*cont.*)
  unstoppable growth of 114–15
global South
  austerity imposed upon 116–17
  blurring of North–South
    distinctions 121–3
  casualization/deregulation of work
    in 114–15
  and community development,
    empowerment, participation,
    ownership 65
  computer availability for 91
  connectivity with 16
  and creation of global precariat
    13–14
  effect of financial crisis on 119–20
  effect of network capitalism on 26
  and humanitarian technologies 166
  informal sector, black market,
    shadow economy in 117–21,
    131
  mobile telephony in 133–4, 151
  NGO invasion of 17, 40
  North–South interface 2–3, 11, 16,
    23, 27, 32
  and post-humanitarianism 10
  and the post-social 12
  post-social context 184
  precarity in 116–17
  role of disasters in 2–3
  as site of disaggregated biopolitics
    of permanent emergency 115
  smart technologies in 133, 151
  structural adjustment in 82
  as testbed for new technologies 11
  transformation through electronic
    globalization 137
  use of term 193n2
globalism, globalization 1, 15, 93,
    129–30, 193–4n1
  second-wave 137
  third-wave 136–7
Goodhart, D. 29
Google 135–6, 136, 156, 201n8
Google Earth 148, 149
Google Earth Outreach 148
Gorz, A. 80
GPS *see* Global Positioning System

Graham, S. 131
Green, D. 179
Gregory, D. 7
Grinstead, N. 150
ground friction 7, 24, 133, 147, 191,
    1933

Haidt, J. 113
Haiti 156
Halpern, O. 4, 7, 29, 35, 37, 116,
    177
Hammerstad, A. 6, 63
Hanchard, D. 145, 146
Harman, G. 130
Harper, L. 112–13
Harrell-Bond, B. 64
Harris, C. 151
Haslam, N. 113
Hayek, F. 20–1, 179
Hayes, T. 145
Hayles, K. 37, 129
Healy, S. 7
Heath, A. 121
HERR 100, 146
Hewitt, K. 26
Hewitt, V. 87
Hobsbawm, E. 31
Hogan, J. R. 96
Holling, C. S. 3, 114
Homer-Dixon, T. 3, 36
Homo economicus 21, 101, 126
Homo iscius 21–2, 27, 29, 40, 41, 72,
    82, 89, 101, 107, 111, 126,
    144, 155, 158, 174, 179, 194n5
Honan, M. 135
Horkheimer, M. 34
Horn of Africa 60
Hosein, G. 128, 164
Howell, A. 105, 109
HRE *see* Human Rights Education
Huawei 139
Human Rights Education (HRE) 149
humanitarian innovation/design 159
  appeal of 175
  attentiveness/care ethos 170, 171
  background 160–1
  and blurring of interface between
    economy and disaster 166

cash transfer 161–2, 167
community/NGO relationship
   169–70
continues work of fantastic
   invasion 167
and direct action of the object 165
drone delivery 163
emergency shelter 162
energy 162–3
and ethical irony 174–6
evidence-based approach 171–2
fetishizing of technologies 164–5
individuation 164
and innovative evidence-based
   responses to global South
   environments 166
measuring success of 171–6
and medicalization of social
   problems 172–4
objects/technologies of 159, 161–4
ontopolitics/centrality of 188
operational logic of 176–7
and remote management/user
   surveillance 166
sanitation 163
secondary market in global North
   175–6
self-help apps 163
therapeutic foods 163, 167
and transience of intermediate
   technology 169–70
water 162
humanitarian neophilia 160
humanitarianism
and digital decentralizing of power
   156–7
direct action 14, 58–62, 98, 168,
   197n4 & n9
Ground Rules in conflict zones
   97–8
loss of flexibility/spontaneity 98–9
and the refugee camp 64–6
sans-frontières 93
*Humanitarianism in the Network
   Age* report (UNOCHA, 2013)
   157–8
Hungary 42
Huws, U. 123

hyper-political age 150–1

Iazzolino, G. 161
ICRC *see* International Committee of
   the Red Cross
identity 65, 79, 87, 140, 164, 166,
   169
IFRC 158
IKEA 162
illiberal regimes 1–2
ImageSat International 149
immigration 6, 31–2, 194n3
Indian Space Research Organisation
   201n5
indignant objects
   background 164–6
   connected logic 166–8
   user communities 168–71
informal/shadow economy 137
   positive orientation towards 126
   and recycling poverty 123–6
   rehabilitation of 126
   spread from South to North
      121–3
information network 15, 82–4
infrastructure 131–2, 161, 162–3,
   185–6, 200n1
integrated mission 100–1
Internally Displaced Persons (IDPs)
   146
International Committee of the Red
   Cross (ICRC) 103–4
International Federation of Red Cross
   and Red Crescent Societies
   (IFRC) 158
International Monetary Fund (IMF)
   116
Internaut 8, 193n4
Internet Governance Forum (2013)
   128
interventionary cosmopolitanism 59,
   197n2
Iran 42
Iraq 13, 93, 99, 120, 148, 196n4
IRIN 110, 111
'Irrelevance of Development Studies'
   (Edwards) 89
ISSUE journal 45

233

Jackson, T. 127
Jacobsen, K. L. 11, 128, 164
James, C. L. R. 9
Japan 40, 42
Jarmolowski, M. 110
Jaspars, S. 55, 57, 77, 80, 84, 92, 150, 168, 172
Johnson, C. G. 166
Johnson, D. H. 44
Jones, O. 25
Joseph, J. 161
Joshi, D. 122, 184
JRC *see* EU Joint Research Centre
Juba 95, 96, 98
Jutting, J. P. 117, 118

Kahn, C. 146
Kaldor, M. 100
Kaplan, R. D. 1
Karadawi, A. 64
Karim, A. 97–8
Keen, D. 60
Kemper, T. 146
Kent, R. 114
Kenya 2, 30, 105, 134, 135, 138–9, 145
Khartoum 47–8, 54, 57, 60, 76, 94, 196n5, 199n1
Kibera slum (Kenya) 138
Klein, N. 114
knowledge
  area-specific 58
  autonomy 34
  and cybernetics 37
  and data 3, 8–10, 11, 21, 51, 76, 92, 115, 151, 178
  fantastic invasion, project form, information exchange 80, 81, 82, 83, 84, 87
  fields of 50–1
  humanitarian 55, 76, 78
  and ignorance of the world 21
  jobs/services 39
  local 88–90, 127
  professional 90
  recipient 172
  social 62, 144
Kosovo 145

Kranz, O. 146
Kristof, N. 116

Lafontaine, C. 67
Laiglesia, J. T. de 117
Lakoff, A. 144
Lancet 172–3
Lang, S. 146
Large, D. 57
Latin America 31
Latour, B. 27, 34, 80, 164, 170, 187–8
Lavers, C. 149
Lavinas, L. 126, 162, 166
Leader, N. 100
LeBaron, G. 26, 114–15, 123, 137, 170, 181
Lee, C. K. 119
Lesczynski, A. 132
Lévi-Strauss, C. 49
Levine, I. 97
Lewis, M. 5
liberal interventionism 113
liberation movements 30–1
Libya 150, 196n4
LifeStraw 162, 168, 170, 171
Live Aid 59
livelihood
  and coping systems 89
  as ever-changing project 79–80
  and the network 79
  NGOs' regime 80–1
  operationalizing 85–7
  relief-to-development continuum 84–5
  replaced by resilience 92
  and self-reliance 86
London 42
Lugard, Lord 44
Luxembourg, Rosa 195n6

McAfee, A. 39
McBain, S. 7
Mckinsey Global Institute 132
Macmichael, H. A. 44
McNamara, Tim 156
Macrae, J. 100
Mahmoud, F. 41–2

INDEX

Mai Wurno, Mohammed Bello 196n7
Maiurno
  arrival 46–7
  communication by letter-writing 48–9
  data collection / excursions 51–3
  decline in peasant-based subsistence economy 53–5
  location 47–8, 196n5–n7
  return to 55–6, 197–8n10
  structural method of fieldwork in 49–51
Malcolm, J. 163
MapAction 148
Marcuse, H. 2, 6, 19, 28, 30, 31, 33, 34–5, 40, 41, 43–4, 59, 66, 69, 130, 195n1
Marvin, S. 131
Marx, K. 34
Marxism 19, 27, 45, 71
'Marxism and Development Sociology: Interpreting the Impasse' (Booth) 71
May '68 movement 12, 18–20, 27, 28, 29, 33, 44, 45, 58–9, 64, 81, 90, 130, 165, 174, 194n4, 196n2
  see also university campus, and student campaigns
Meagher, K. 115, 117, 123, 124, 125, 137
Médecins Sans Frontières (MSF) 59
medicalized interventions 172–4
Meier, P. 10, 77, 143, 153, 154, 156, 157, 182
MERIP *see* Middle East Research and Information Project
Mexico City 42
Meyers, E. 6, 32
Middle East 104, 120–1, 160
Middle East Report 45
Middle East Research and Information Project (MERIP) 45
Mills, C. Wright 2, 28, 33, 40
Mind, Society and Behaviour report (World Bank, 2015) 89, 179–82, 185

Minimum Operational Residential Security Standards (MORSS) 103
Minimum Operational Security Standards (MOSS) 103
Minton, A. 133
Mishra, P. 7, 140, 190
MIT 90
mobile banking 161–2
mobile telephony 11, 16, 91, 98, 104, 113, 115, 125, 130, 155
mobility differential 25–6
modernity 188
Monk, D. B. 133
Montevideo 42
Montgomery, C. 99
Morozov, E. 4, 178
Morrow, M. 164
MORSS *see* Minimum Operational Residential Security Standards
MOSS *see* Minimum Operational Security Standards
Mozambique 85, 86, 196–7n2
MSF *see* Médecins Sans Frontières
MSR *see* Munich Security Report
MTN 139
Muhren, W. J. 152, 155
Munck, R. 115
Munich Security Report (MSR, 2017) 1, 3, 93, 190

NACLA *see* North American Congress on Latin America
NACLA Report on the Americas 45
natural disasters 63, 119
Naughton, J. 135
negative dialect 33–5, 195n5-n7
Negroponte, N. 90–1, 157
neo-fascism 33–4
Neocleous, M. 136
neoliberalism 20–1, 25, 65, 75, 181, 194n5
  Hayekian 126
  and realizing capital's potential 84
  and security 198n12
  shaping of 33
Nepal 145
network capitalism 29, 79

network capitalism (*cont.*)
  change in employment 79–80
  and death of the social 22–4
  development of 16
  entrepreneur as networker/celebrity 23–4
  and global precariat 140
  inequality, youth poverty, indebtedness 23
  mobility vs immobility 24–6
  and progressive neoliberalism 20–1
  as site of permanent emergency 23
  terminology 16, 194n2
  see also new economy
network society 15
new economy 16, 69–70, 79
  see also network capitalism
New Left 2, 12, 27, 86
  break with Victorian Marxism 33
  and decolonization 29–31
  development of 28–9
  dismissal of working class 40, 196n2
  and diversity of Third World 30
  embracing of Maoism / Khmer Rouge 194n1
  negative dialectic 33–5, 195n5
  and the new anthropology 43–6
  peasantry as force for revolution / national liberation 30–1
  questioning of internationalism 53–4
  resisting entropy 35–8
  revolution predicated on circulation 29
  role of intelligentsia as 'lever of history' 39–40
  solidarity transformed in ethics of irony 174–6
  student unrest 40–3
  and world as space of political possibility 31
  and world revolution 91
New Left Review (NLR) 29, 42, 43
New York 42
New Zealand 6, 31
Nicaragua 194n2
NLR see *New Left Review*
non-governmental organizations (NGOs) 13, 17, 74, 75
  attacks on 'charity' of aid agencies 197n4
  autonomy/independence of 59–61, 85
  and communication/information organization 82–4
  and computer learning 90–2
  direct humanitarian action 58–62, 197n9
  donor control over 85
  early adopters of project form 80–1
  expansion of 58, 196–7n2
  exploitation by 62
  failures of remedial interventions 117
  fantastic invasion of 59–78, 83–4, 85, 88, 153–4
  and foreign policy goals 100–1
  fortified aid compounds 94–104
  geospatial data given to 146
  increased involvement of 75
  introduction of log frame analysis 85
  and launching of careers 88
  and local knowledge 88–90
  and making communities visible 85–6
  mobility differential 61–2
  mobility of 88
  participatory process 89–90
  as pioneer of postmodernist relations / governance structures 81–2
  political aspects 59, 60–1, 197n7 & n8
  and social reproduction 84–7
  in Sudan 55, 57–8
  telex communication 197n7
Nordstrom, C. 117
North American Congress on Latin America (NACLA) 45
North–South interface 2–3, 11, 106, 107, 178, 190, 193n2
Nussbaum, B. 159, 166
nutrition, medicalization of 172–4
Nyst, C. 128, 164

O'Brien, J. 53, 70

O'Malley, P. 109
O'Reilly, T. 143, 152
OAU *see* Organisation of African Union
ODI *see* Overseas Development Institute
OECD *see* Organisation for Economic Co-operation and Development
off-grid services 135, 140, 141, 158, 159, 167–8
Office of the United Nations Security Coordinator (UNSECOORD) 103, 104
Okinawa 40
One Dimensional Man (Marcuse) 195n1
One Laptop per Child 91
Open Knowledge Foundation 156
Operational Lifeline Sudan (OLS) 97–8, 105, 200n5
Operational Security Management in Violent Environments (Van Brabant) 102
Orange 156
Orangi Town slum (Karachi) 125
Organisation for Economic Co-operation and Development (OECD) 118, 119, 120
Organisation of African Union (OAU) 63
Overseas Development Institute (ODI) 162
Oxfam 57, 60, 61, 73, 74, 76, 118, 194n2, 197n5 & n7, 198n14, 201n10

Packard, V. 41
Page, T. 6, 32
Palen, L. 152, 153
Parks, L. 148, 149
patriarchy 16, 19
peasants 2, 30–1, 44
Pelosky, R. J. 154
personal freedom 23–4
Petti, A. 24, 112
Piketty, T. 118

Pirlot, A. 128
Planet of Slums (Davis) 117
Plumpy Nut 163, 167
Poell, T. 132
Poland 42
political pessimism 5
political push-back 7
population 22
Portishead Coastal Radio 197n7
post-Fordism 12, 15, 22, 54, 79, 167, 194n5
post-humanism 130–1
post-humanitarianism 8–10, 12
  difficulties visualizing progress 141–2
  emergence of 10, 116
  and jobs on the ground 168
  and post-humanism 130–1
  remote sensing / mobile data informatics as conditions of existence 143–4
  and self-acting community 84
  and streamlining access to off-grid services 167–8
  technologies 167
post-social world 10
  anticipation of 87, 92
  and capitalism 125, 137, 161, 168–9
  career/work structures 16–17, 168, 184
  and the community 65
  as expansive, ruined, wild landscape 189
  exposure to 74
  and fantastic invasion 13, 80, 87
  freedom of the market in 12
  and global precarity 115
  and global South 184
  in the global South 139
  and humanitarian innovation 161
  and modern slavery 137
  optimism concerning 79
  reaching/connecting across mobility barriers 121
  resilience in 23, 125
  and security 66, 72
  and smart technologies 14, 129, 170

237

post-social world (*cont.*)
  survival strategies 80, 87, 159
post-traumatic stress disorder (PTSD) 105, 110
postmodernism 12, 65–6
poverty 82, 86, 118
  and active unemployment 118–20
  bucolic depiction of 127
  as a cognitive tax 181
  computer learning 90–2
  decline in 119, 200n2
  effect of financial crisis on 119–20
  in Kenya 138–9
  local knowledge 88–90
  and nutrition / food aid 172–4
  as personal experience 180–1
  programming 87–92
  recycling 123–6
  redefinition in terms of 'bandwidth' 180
  and redistribution of resources 180
  relevance of Middle East urbicidal wars 120–1
  surviving 115–16
  and use of crisis informatics 153
Prahalad, C. K. 125
precarity 26
  see also global precariat
Prendergast, J. 61, 149
Prins, E. 147
progressive neoliberalism 20–2, 126–7
project form
  action-orientation of 83, 168
  and capacity-building 83–4, 199n2
  community action 83–4
  development 79–81
  emergence of network capitalism 83
  fantastic invasion's notions of 83–4, 159–60
  information flow 81–2, 83–4
  intellectual infrastructure 82–3
  as mechanism of exploitation 87–92
  move beyond 127
  and NGOs 81
  remote management of 99
  repurposing of 199n3
  and social reproduction 84–7
projective city 80
PTSD *see* post-traumatic stress disorder

Rabinow, P. 4, 11, 22, 69, 131, 181
Radio Dabanga 149
Rahebi, M.-A. 70
Ramalingam, B. 179, 183
Read, J. 190
Red Crescent Society 158
Redfield, P. 84, 133, 161, 162, 165, 166, 168, 169, 170, 171, 175
RedR 102
Refugee Convention of the Organisation of African Union (OAU) 63
refugees
  biometric registration of 164
  camps for 58, 63–6, 145–6, 147–51, 198n11
  record numbers of 120–1
  remote sensing/management of 145–6, 147–51
Reid, J. 3, 22, 51, 124, 129, 152, 189
Relief and Rehabilitation Commission (RRC) 76
remote sensing
  application of 144–5
  before-and-after satellite photography 149
  commercial contracts 145, 200n1, 201n3
  crisis on the ground in Darfur 148, 149–50
  and humanitarian organizations 145–7
  increasing use of 144
  mapping initiatives 148–9
  military use 144–5, 200n1, 201n3
  and normalization of absence / decreasing reliance on ground truth 147
  public advocacy of 148–9
  and refugee camps 146–7, 150–1
  seeing is believing 147–51
  viewed in relationship to its environment 158

remoteness 7–8
resilience 126
  anticipation of term 63, 76
  and 'authentic life' 3
  current regime 65
  demands for 104
  ecological 3, 63, 194n6
  ecology-based 3
  enabling through crisis informatics 154–5
  and entropy 36
  erection of 22–3
  and first-order cybernetics 37
  focus on 80
  livelihood regime replaced by 92
  and management of ecological systems 63
  personal 105
  photographic hopefulness 142
  rise of 114, 125
  training 13, 111
RESPOND 145, 146, 201n5
Review of African Political Economy (ROAPE) 45
revolution
  based on circulation 29
  precursors of 29–32
revolutionary optimism 5
revolutionary struggle 2
Richmond, O. 93, 106
risk aversion 7, 12, 58, 60–1, 147
ROAPE see *Review of African Political Economy*
Robins, S. 163, 175
Robinson, Mary 107–8
Roitman, J. 117
Rome 42
Rose, N. 11, 22, 23, 65, 79
Roser, M. 121
Rostow, W. W. 84, 132
Rouvroy, A. 9, 28, 37, 51, 52, 64, 70, 183
Ruel, M. T. 172, 173
Russia 42
Rwanda 102
Rydjeski, D. 77

Safaricom 139

San Francisco 42
Save the Children Fund 75, 198n14
SBTF see Standby Task Force
School of Oriental and African Studies (SOAS) 46
Schumacher, E. F. 81–3
Scott-Smith, T. 160, 163, 165, 167
security
  of aid workers 100–4
  and biopolitics 66
  centralizing tendency 102, 103–4
  and circulation/mobility 5, 6, 24, 25–6, 47, 66–8
  comprehensive protocols 60–1
  disciplinary mechanisms 67–8
  and fortified aid compounds 94–104, 147
  governmental urge to leave nothing hidden/undisclosed 66
  and Ground Rules 97–8
  increased 55–6
  and information-gathering 98–9
  latest trends/events 1
  liberal conception 4, 13, 29, 58, 66, 198n12
  and liberalism 198n12
  and the media 32
  and need for inner resilience/strength 109–11
  and perception of aid work as dangerous 99–102
  and period of decolonization 29–32
  and personal freedom 23, 97
  personalizing 105–11
  as place of uncertainty/complexity 29
  praise for 98
  and relative openness / shared connectivity 32
  remote management techniques 99
  restrictions/bureaucracy as result of 98–9
  spatial/temporal asymmetry between mobility and immobility 140–1
  system-based approach 103–4
  training framework 102–3, 105–6

INDEX

security (*cont.*)
  and welfare-Fordism 16–18, 22, 23, 79
security training
  Basic/Advanced modules 106–7
  enjoyment of experience 107
  in the field 13, 102–3, 105–6
  generic characteristics 106
  importance of nurturing inner self 109–11
  institutionalization/governmentalization of 107–8
  in practice 108–9
Seddon, D. 117, 140
Seekers of Truth and Justice (STJ) 198n13
Sen, A. 75
Setel, P. W. 164
Shoham, J. 84
Silicon Valley 135, 140, 152, 156
Simondon, G. 37
Skinner, D. 133
skivers and strivers 25
slaves, slavery 9, 137, 165
Slim, H. 64, 164, 167
Sloterdijk, P. 5, 29, 48, 104, 187
slums 125, 138, 142
Small is Beautiful: A Study of Economics as if People Mattered (Schumacher) 81–2
smart technology 131
  and humanitarian innovation 141
  hype/fetishism surrounding 150
  levelling downwards 14, 128, 131–3, 134, 157, 159, 183
  in urban development 132–3
Smirl, L. 112
Smith, J. E. 122, 123
SOAS *see* School of Oriental and African Studies
social
  death of 22–4
  exclusion 25, 75
  media 135, 136, 156, 183–4, 201n8
  reproduction 84–7, 116, 184–6
Solar Aid 162–3
solar power 162–3
Somalia 99, 102

Sorensen, J. S. 7
South Korea 40
South Sudan 96–9, 105, 112
Southern Africa 31
sovereignty 12, 27, 29, 32–3, 50
Soviet Union 43
special economic zones (SEZs) 137
Spencer, D. 21, 105, 112, 126, 133, 165, 183
Spiz, P. 75
SPLA *see* Sudan People's Liberation Army
Spreeuwenberg, K. 132
Srnicek, N. 8, 157
Standby Task Force (SBTF) 201n10
steersman analogy 68–70
Stiegler, B. 17, 21, 23, 29, 129, 140, 170, 192
STJ *see* Seekers of Truth and Justice
Stoddard, A. 99–100, 107
Stover, E. 76
Streeck, W. 114, 121
Street, A. 160, 175
structural adjustment programmes 116–17, 200n1
structuralism 49–51
Sudan 201n10
  aid compounds 94–9
  airstrips in 97, 200n4
  appearance of complexity thinking in 70–7
  being there 51–3
  data collection 51–2, 76
  dignity of distance 46–9
  discussion topics 53
  dissolution of peasant agriculture 54–5
  emerging divisions/inequality in 53–4
  famine in 72–4, 76
  fieldwork 46–56
  fortified aid compounds in 94–6
  friendships made/continued 53
  industrialization in 80, 199n1
  infrastructure 54–5
  NGO invasion of 55, 57–8
  outcomes 53–5
  participant observation 52–3

INDEX

retrospective study 55–6
return to 55–6, 70, 197–8n10
as site of fantastic invasion 57, 196n1
structural method 49–51, 196n8
telecommunications in 48, 60, 197n7
transition from Fordism to post-Fordism 54
Sudan People's Liberation Army (SPLA) 97
Sudan the Roots of Famine (Cater) 72–3, 76
Sudanese Bourgeoisie, The (Mahmoud) 70–1
Sulik, J. J. 146, 147, 149
Sunstein, C. 183
Syria 1, 120, 201n10

al Tahir, Abu Bakr Mohammed 47, 196n7
Taiwan 40
Taplin, J. 90, 178
TaskRabbit 201n8
TATA 139
Taylor, L. 128
techno-pastoral aesthetic 126–8, 131, 140–2, 174
technological determinism 165–6
technoscience 34, 132, 141–2, 147, 157, 159, 161, 189–90, 191
Terranova, T. 132
terrorism 1, 174
Tethered Goat, The (Winer) 197n8
Thaler, R. 183
Thinking and Working 199n7
Third World 2, 71
  diversity of 30
  and economic catch-up with the West 82
  and ideology 187
  and informal economy 124–5
  infrastructure 131–2
  progressive ethos of 32, 194–5n4
  as progressive geopolitical entity 30
  and struggle for national liberation 30–1, 43

top-down/state-led modernization in 81–2
Tiller, S. 7
To Kill a Mockingbird (Lee) 112–13
Tofler, A. 141, 142
Toronto 4
town
  circulatory potential 4
  as closed interactive milieu 4–5
  data-based urbanism 4
Truman Doctrine 31
Trump, Donald 7, 184
TUC 23
Tunis 42
Turkey 40
Turner, F. 19, 20, 28, 36, 37, 90, 152
Tvedt, T. 57, 97
Twitter 156, 201n8

Uber 201n8
Uganda 196n5
UN *see* United Nations
UN-Habitat 116, 117, 118
UNASF *see* United Nations Advanced Security in the Field
UNBSF *see* United Nations Basic Security in the Field
UNDP *see* United Nations Development Programme
UNDSS *see* United Nations Department of Safety and Security
UNECA *see* United Nations Economic Commission for Africa
UNGP *see* United Nations Global Pulse
UNHCR *see* United Nations Refugee Agency
United Nations (UN) 26, 61, 99, 100, 103, 119, 124
United Nations Advanced Security in the Field (UNASF) 106
United Nations Basic Security in the Field (UNBFS) 106, 108, 109, 110
United Nations Department for Humanitarian Affairs 72

INDEX

United Nations Department of Safety and Security (UNDSS) 104
United Nations Development Programme (UNDP) 84, 127
United Nations Economic Commission for Africa (UNECA) 138
United Nations Food and Agriculture Organization (FAO) 75
United Nations Global Pulse (UNGP) project 66, 153, 154, 182
United Nations Global Report on Human Settlements (2003) 117–18
United Nations Mission in Sudan (UNMIS) 94
United Nations Office for the Coordination of Humanitarian Affairs (UNOCHA) 146, 153, 157–8
United Nations Operational Applications Programme (UNOSAT) 145, 146
United Nations Refugee Agency (UNHCR) 6, 63–4, 94, 95, 97, 120–1, 145, 146, 148, 162, 164
university campus
  censorship/'no-platforming' in 112–13
  as liberated space 113
  and student campaigns 41–3, 44
  see also May '68 movement
University of Khartoum 46, 70
UNMIS see United Nations Mission in Sudan
UNOCHA see United Nations Office for the Coordination of Humanitarian Affairs
UNOSAT see United Nations Operational Applications Programme
UNSECOORD see Office of the United Nations Security Coordinator
Urry, 73
US Holocaust Memorial Museum (USHMM) 148

USAID 76, 172, 201n10
Usborne, D. 7
USHMM see US Holocaust Memorial Museum

Van Brabant, K. 102, 107, 108, 109
Van de Walle, B. 152, 155
Verjee, F. 144, 145, 146
Vestergaard Frandsen 168
Vietnam 2, 30, 32, 42, 43, 59, 194n2
Virilio, P. 24
Vodafone 156
Volunteer Technical Communities (VTCs) 156, 157–8
VTCs see Volunteer Technical Communities

Waal, A. de 75, 77
Walker, P. 57
Walton, J. 117, 140
War on Terror 189
Washington Times 152
Weber, M. 17
WEF (World Economic Forum) 118, 125
Weizman, E. 24, 113
welfare-Fordism 16–17, 131
  corporate manager 18
  factory-owning entrepreneur 18
  as focus of anti-capitalist critique 18–20
  and security 16–18, 22, 23, 79
Western decline 1–2
Wiener, N. 5, 21, 35–6, 68
Wiley, D. 45
Winer, Nicholas 197n8
Wisner, B. 63
Wolpe, H. 86
women 86–7, 118, 138
Wood, D. 8
Woodis, J. 30
world alienation 7, 8, 10, 113, 190–1
World Bank 9, 21, 101, 111, 116, 117, 119, 133, 162, 164, 167, 179, 179–81, 183, 185–6, 186
World Development Report (2015) 179

242

World Disasters Report (IFRC 2013) 158

youth unemployment 125
Yuma Proving Ground 136

Zook, M. 152
Zuckerberg, M. 26, 135, 136
Zyck, S. A. 114